GPRS and 3G Wireless Applications

Professional Developer's Guide

GPRS and 3G Wireless Applications

Christoffer Andersson

Wiley Computer Publishing

John Wiley & Sons, Inc.

NEW YORK • CHICHESTER • WEINHEIM • BRISBANE • SINGAPORE • TORONTO

Publisher: Robert Ipsen
Editor: Carol Long
Managing Editor: Angela Smith
Text Design & Composition: D&G Limited, LLC

This book is printed on acid-free paper. ∞

Published by John Wiley & Sons, Inc.

Published simultaneously in Canada.

Library of Congress Cataloging-in-Publication Data:

Andersson, Christoffer
GPRS and 3G wireless applications: professional developer's guide/Christoffer Andersson.
p. cm.
Includes bibliographical references and index.
ISBN: 0-471-41405-0 (cloth: alk. paper)
1. Wireless communication systems. 2. Mobile communication systems. I. Title.

TK5103.2.A53 2001
621.382—dc21 2001017825

Printed in the United States of America.

10 9 8 7 6 5 4 3 2

Professional Developer's Guide Series

Other titles in the series:

Advanced Palm Programming by Steve Mann and Ray Rischpater, ISBN 0-471-39087-9

WAP Servlets by John L. Cook, III, ISBN 0-471-39307-X

Java 2 Micro Edition, by Eric Giguere. ISBN 0-471-39065-8

Contents

Acknowledgments

In a book this packed with knowledge it is very hard to trace the source of many of the ideas and conclusions. I have read many documents and books in this area and the digested content has been one source of input for this book. None mentioned and none forgotten, I hope that those other authors in the industry can feel that they contributed to this book by documenting their ideas and giving me a head start. The mobile Internet industry is forming and we all need to share our knowledge to take it all to the next level. This thinking is especially used within the Mobile Applications Initiative (MAI), where the aim from the start has been to grow the knowledge among developers and ensure that wireless applications function smoothly when deployed. With over 30 centers and a dedicated global organization, I would not dare start pointing at individual contributors to express my thanks. This MAI knowledge network has built a large portion of the knowledge that I have today and many of those contributors would have been as appropriate to write this book.

In addition, the support from Ericsson has been amazing. Some of the company's leading executives and experts have helped a lot in developing and reviewing the material. I especially want to mention Mats Dahlin and Göran Skyttvall, who have been initiators and stong drivers along the way. The many junctions where decisions were needed were promptly and decisively handled.

Reviewing a book like this is an enormous task and requires deep knowledge in a wide range of areas. Tons of thanks go to Gwenn Larson, Luis Barriga, Larry Wood, Daniel Freeman, Per Lindtorp, Peter Lowten, Göran Svennarp, Hans Olsson, Bo Johansson, Stefan Berggren, Anna Hultman, Jonas Lindh, Ari Jouppila, and Derek Sellin for helping me lift their deep knowledge to the silver plate that I wanted it served on for the readers.

Getting the courage to initiate the writing of a book is a big step. Martin Gutierrez planted the seed the first day I met him, and that thought never left my mind.

In my working life I have always been fortunate to meet interesting and helpful people. This started already when graduating from Luleå University. Staffan Johansson from Ericsson Erisoft and Svante Carlsson from the University contributed so much more to my personal development than they ever understood.

Finally, I strongly believe that growing up with the loving family I did and meeting a wonderful person like Malin is what makes the world a sunny place, even in the dark Swedish winter.

Introduction

The mobile Internet—the marriage between today's Internet and the increasing urge for mobility—is about to take off. Independent sources all estimate incomprehensible markets for these new services and applications. With the mobile Internet, there is a screaming need for applications (in other words, things to do with the new technologies). In other words, we dearly need the software developer and Internet communities in order to use their expertise for developing software and content. These developers know about programming and Web design but do not know much about the wireless technologies of today and of the future. This book provides a broad guide to the networks, devices, and other items that surround the applications (such as positioning, security, and how to deploy applications in the field).

This book is primarily intended for those who know about software development and who want to do it for wireless networks and devices, but this book should also appeal to anyone who is interested in these new and exciting topics. Although some chapters assume knowledge about development, you can read most of them as a technician's guide to wireless technologies.

The telecommunications industry has traditionally been a walled garden where no one but the insiders knows much about the technology. This situation is rapidly changing, however, now that players are entering the arena and are contributing to the growth of this new and exciting market. This book is an invitation to you to join the mobile Internet revolution.

Overview of This Book and the Technology

When you are trying to make things work with a new technology, you need to combine many forces. Infrastructure and handset manufacturers such as Nokia, Ericsson, Motorola, and so on and the wireless operators Vodaphone/Airtouch, Sonera, and AT&T have set the groundwork for the mobile Internet. They are all committed to deploying *General Packet Radio Services* (GPRS) and *third-generation* (3G) networks. Now, these players feel confident that their parts of this new market (technologies such as WAP, Bluetooth, GPRS, and 3G) will be in place and that it will all work. This situation raises a need to mobilize other players to contribute: the applications (mail, games, chat, Java applets, and so on) and content developers (Web, the Wireless Access Protocol or WAP, and so on). These players are dying to move into this space, and there is a need to gain the knowledge about these wireless technologies and how it affects their products. Today, we see too many developers ignoring these properties, failing to make the applications tolerant to scenarios of going through tunnels and being in areas of weak signal strengths. Every developer who creates applications (WAP or others) for GPRS and 3G wireless systems needs a handbook that clearly illustrates how you can overcome difficulties and leverage the new possibilities of the mobile Internet.

Having worked with software and Internet developers over the past couple of years (helping them optimize their applications for wireless networks), I find that many questions occur over and over again. I did not anticipate many of these questions when I was designing the 3G systems. These questions could, for instance, be as follows:

- How do I make my application cope with difficult radio conditions, such as going through a tunnel?
- How can my application access the features of the network, such as signal strength and the *Quality of Service* (QoS) that is used?
- How will my application typically be implemented in a mobile operator's network now and in the future?
- What is required if I want to add location dependence to my product?

At first, I found it difficult to find answers to these questions, because most resources do not always convey the big picture, and lots of information on the

Internet is even incorrect. The books that are available to date are about 300–400 pages long for each technology described. The problem arises when the average developer needs to know a bit about GPRS, *Enhanced Datarates for GSM and TDMA Evolution* (EDGE), *Wideband Code Division Multiple Access* (WCDMA), cdma2000, WAP, Bluetooth, EPOC, and kJava (to name just a few). Reading a book about each of maybe 10 to 20 technologies can be pretty overwhelming just to get started. My approach is to gather the parts of these technologies that are relevant to application developers, which means 20 to 30 pages about the main technologies and a bit less about the peripheral ones. I then complement this information with concrete advice about how to create successful applications that are optimized in order to work on wireless networks. This book is very technology focused, and I have left most of the business aspects until the last chapter. In the other chapters, I only mention a topic if it affects the technology decisions. One example is the discussion about the number of subscribers for the different technologies.

In essence, after reading this book the reader should have a very good understanding of the different mobile technologies and how they are related. This knowledge should be enough to make the necessary initial decisions in the development process, such as choosing the application architecture (for example, putting software on the device or just using a browser). Once the development is underway, the book will provide valuable advice about how to get the most from the mobile Internet and how to overcome some of its inherent difficulties. Once the desired platform is chosen, there are other good books that complement this one and that go deeper into the actual coding of the product. While great books exist about the programming languages and the operating systems, it is harder to find a book about mobile networks. This situation partly depends on the rapid pace of standardization, where small changes frequently occur. The standards are sometimes tricky to find and to understand, but the small guide at the end of this chapter should make you confident. When developing things that interact intimately with the networks and devices, you should always use the standards as the authoritative source of information. In the text, I make reference to the document numbers of the standards that provide a deeper discussion about the topic.

How This Book Is Organized

This book is divided into three parts:

Part One, “The Mobile Networks,” explains how the wireless networks work and how they complement each other. In the beginning, we explain the basic terminology in order for people who are not from the tele-communications industry to understand the acronyms (and there are many of those). The descriptions are consistently from the developer's point of view, rather than from the point of view of the mobile operator or the handset vendor. The application developer who wants to succeed needs to know the basics of the networks and how the properties look at the application level. After going through this part, you will be able to talk fluently with the people in the business (who have forgotten how little they knew in the beginning). This knowledge makes it possible to ask the right questions and to start discussions with potential partners. The first part includes Chapters 1 through 5.

Part Two, "Optimizing the Transmission," then talks about how all of these items affect the applications. The starting point is the existing datacom networks and the most commonly used protocols. This section analyzes shortcomings as well as opportunities and compares them with newer additions (such as WAP). This part also includes my experiences from working with developers over the years, helping them optimize applications. The most common mistakes appear over and over again when applications are tested, and we hope you will avoid the mistakes that others have made. The information in this part is incredibly valuable to any developer, no matter whether a new application is developed or whether an old one is optimized for wireless. Many of the companies with which I have worked have also found that the optimized application now functions better on fixed networks such as *Local Area Networks* (LANs). The second part includes Chapters 6 through 8.

Part Three, "Applications and Their Environments," digs deeper into the actual applications and other components that interact directly with them. The chapters in this part are generally shorter, and we cover more technologies in less space. The average developer maybe gets involved with half of the technologies; therefore, a more superficial coverage level is appropriate. Some of the chapters also involve topics where standardization is still underway, such as positioning and kJava. As with the rest of the book, we try to paint a picture of what you can do now and when you can expect standardization. The last chapter includes some of the experiences that I have had on the business side and what it takes to succeed all the way. The third part includes Chapters 9 through 15.

You can read the parts independently (the same goes for the individual chapters).

Chapter 1, "Basic Concepts," sets the stage and examines some of the basic concepts. Here, I also go through some of the naming conventions that I use in order to clarify how I see the meanings of the concepts that I use frequently. People who are familiar with cells, base stations, and *Time Division Duplex* (TDD) should probably just skim through this chapter in order to catch the definitions of words such as *application* and *fixed Internet.*

Chapter 2, "The Mobile Evolution," draws the big picture of the mobile systems of yesterday, today, and tomorrow. This chapter provides a better understanding of why we have several different *second-generation* (2G) networks and why it has been hard to get one unified 3G standard. We describe the steps for each migration path into 3G and provide an understanding of the expected timelines as well as the challenges that operators face.

Chapter 3, "GPRS—Wireless Packet Data" explains how *General Packet Radio Services* (GPRS) works and why it is such an important part of the mobile Internet. Although this chapter provides a description of a specific system, it also covers many generic aspects that you can apply to other wireless packet data networks. Despite the existence of a specific chapter concerning devices, we have chosen to include some specific information about the GPRS handsets here. This information includes how to connect other devices to the phone via the R-reference point and how to use AT-commands to access the network properties.

Chapter 4, "3G Wireless Systems," dives into the *third-generation* (3G) wireless systems and how they affect developers. This chapter proved much more difficult to write than originally anticipated. There are four main standards for mobile networks (not counting satellite-borne ones), and the terminology varies a lot depending on whom you ask. There is ongoing work here to harmonize the different standards and to free up spectra in countries where standards were not previously available. Even the last week before the deadline for this book, there were interesting developments—and more things are likely to occur. The main messages and descriptions of the systems should be stable, however.

Chapter 5, "Bluetooth—Cutting the Cord!" describes short-range radio technology and how it greatly complements the *Wide Area Networks* (WANs) in the previous chapters. The focus is again on how Bluetooth can help developers create exciting applications by using the technology. We expect Bluetooth to be extremely pervasive; consequently, it is also mentioned in many of the other chapters. This chapter, however, concentrates on how Bluetooth works and how it appears to the end user.

Chapter 6, "Unwiring the Internet," examines the fixed Internet as we know it and explores its associated protocols (primarily, the *Transmission Control Protocol* or TCP, the *Internet Protocol* or IP, and the *Hypertext Transport Protocol*, or HTTP). We discuss some of the problems that we have found when running these protocols over wireless networks, and we explain what to do in order to cope with these issues. This chapter not only covers the traditional view of TCP/IP over wireless but also expands it to include some issues that we found when running high-speed networks such as Universal Mobile Telecommunication System (UMTS) and retransmitting lost packets.

Chapter 7, "The Wireless Application Protocol," talks about WAP and describes its properties. While most of the existing literature on the topic focuses on the markup language, this description looks more beneath the application's environment. We investigate and compare the underlying protocols that ensure that information is transferred in a robust and efficient way with the TCP/IP protocols of the previous chapter. This information should be a new and refreshing angle on a suite of protocols that we have intensively discussed on a superficial level. We also touch upon the future of WAP and Extensible Hypertext Markup Language (XHTML) at the end.

Chapter 8, "Adapting for Wireless Challenges," includes the bulk of experience that I have gathered from testing dozens of wireless applications and helping developers optimize them. Not only do we describe the most common issues, including interruptions, long latency, and low bandwidth, but we also propose solutions to these problems. This chapter is a must-read for anyone who is developing applications for wireless networks.

Chapter 9, "Application Architectures," describes how you can implement applications on the networks of today and tomorrow. We place emphasis on the upcoming service network architecture, where applications can access the features of the mobile networks by using open *Application Programming Interfaces* (APIs). This functionality will open amazing possibilities for developers who now can use features such as positioning, call control, and charging in order to enhance their products.

Chapter 10, "Mobile Internet Devices," is an introduction to the devices that we can use with the mobile Internet. The most important part is setting the stage for future device constellation and for integrated and divided concepts. In addition, there is plenty of advice that is useful for getting the most from the platform that you use and for saving battery and *central processing unit* (CPU) power. While no one knows what the future of devices will look like, this chapter illustrates some of the changes that are happening and how to adapt one's thinking.

Chapter 11, "Operating Systems and Application Environments," goes one step further than the previous chapter and looks at the items that you can add to devices in order to make application implementation easier. This chapter includes the operating systems Palm OS, Windows CE (Pocket PC), EPOC, and Linux. We also question the role of the operating system and explain the bright future of Java for mobile devices.

Chapter 12, "Security," tries to illustrate the wide concept of security for mobile Internet applications. The reader who is unfamiliar with the concepts will learn the basics of cryptography and how to use it in products. The emphasis is on the big picture and how to weigh the desired security level against the added complexity in order to make the right decisions.

Chapter 13, "Location-Based Services," goes one step further toward describing one of the most important parts of the service network mentioned in Chapter 9: location-based applications. We describe different positioning techniques, both handset based and network based, in order for the reader to realize the consequence of using each one. This description, of course, includes the accuracy for each technology and also some aspects that affect the operator's decision for positioning method. One important aspect that we often forget, however, is the issue of delay that the positioning requests give to an application.

Chapter 14, "Testing the Wireless Applications," adds one of the most important parts of the development process and explains the different parts of the testing process. While the focus is on testing the wireless properties, we also stress the importance of testing for usability and user friendliness. One of the keys to being successful in testing is to work with those people who are experienced and who can give objective feedback. We provide a list of such resources at the end of the chapter.

Chapter 15, "Getting It All Together," summarizes some of the most important nontechnical aspects of creating successful mobile Internet applications. The most obvious part is, of course, how to make money from the application. While we provide no definitive answer to that question, the chapter still provides some valuable leads on where to look. Getting prototypes out early and using the right partners has proven successful for companies with which I have worked. Hopefully this chapter will make you even more successful.

These descriptions only show the topics at a very high level, and due to the many components of the area, the insights will occur in different places for everyone. While the chapters are sufficiently detached in order to enable you to read them in any order, it is highly beneficial to read them through consecutively first. Then, you can use them as independent references when needed.

Keeping Things Objective

My focus when talking to developers is first and foremost to explain what the standard in question says and how they should interpret it. When starting this book, I had the clear objective to keep this view as far-reaching as possible, but sometimes this challenge proved difficult.

The information in this book is very hard to collect for an outsider, and reading the standards will only get you so far. Things such as how implementation issues are commonly solved and what the consumer experience will be regarding technologies that are not on the market are almost impossible for the common man to find out. Because most of my experiences have come from working for Ericsson, it is inevitable that I have researched some topics by using Ericsson material (with loss of objectivity as a result). I also have to grab the opportunity to thank the senior management of Ericsson for letting me share things that are normally not spread outside a limited audience. I am certain that the developer community will find this information helpful.

In those cases where a standard is not finalized, such as positioning, I try to emphasize that we are describing a proprietary solution. I chose to include this information because these things are so important to developers, and the upcoming standard will resemble the existing *software development kits* (SDKs). My general philosophy is that every developer should adhere to the standard if there is one, such as with WAP. When there is no standard ready and migrating from a proprietary solution provides a first-mover advantage, it might be wise to get things going as quickly as possible. Once the standard is available, it is vital for the entire developer community to strive toward using common methods. When designing the application, the developer will find this goal fairly easy because he or she can isolate the API communication and change it easily.

In describing the mobile systems, I explain the standard as far as possible and the Ericsson implementations where there are multiple options. In those cases where there are multiple choices, I mention what the options are and some of the pros and cons associated with each of them. This solution was the best that I could think of, and the reasons for this description are many. First of all, Ericsson has by far the largest market share of 2G systems, and it does not look like it will be less for 3G systems. Second, I have not been part of the system development of Nokia, Lucent, or another system, and getting all of the details would be difficult. The standard leaves many things open, and its description is often too flexible to be easily understood.

When compiling the figures, I have scanned the Internet for high-resolution images of the devices and other components of the future. To me, it appeared that this information was hard to find from all vendors except Nokia and Ericsson. These are the reasons for having most of the device photographs from these companies.

Who Should Read This Book?

The primary audience of this book is the developer community and those who will create the future applications of the mobile Internet. This book is ideal for those who know a bit about software/Web development and who want to get into the wireless field. The book is technical in nature, and those who have a technical background are likely to benefit the most while others can gain a brief overview of the technologies that are involved. This target group benefits from reading the book from start to finish, because it builds the solid wireless competence that is necessary in order to be successful.

Students who want to gain a solid overview of the main driving technologies of the mobile Internet can read those chapters that are appropriate and get a good starting point for further studies. Most likely, the knowledge in this book will be a fundamental part of many software developer and data communications programs at universities.

After having people from the telecommunications industry proofread the content of this book, I found that all of the involved experts have found the book useful. Those who know all about positioning rarely know everything about Bluetooth and all of the other areas that we describe. Wireless experts can probably browse through the first part and hit Part Two and Part Three fairly quickly.

Tools You Will Need

This book provides the foundation for developers of mobile applications in that it describes the components and their interaction. There are, however, very few concrete code samples and example programs. The reason is because you need more than this book to create the application.

While this book provides introductions to the topics illustrated in Figure I.1, we need more in order to go all the way. The aim of this book is to provide a complete guide to the mobile networks and their importance in application development. The number of possible platforms for development is too big, and providing many examples would only limit the usability of this book.

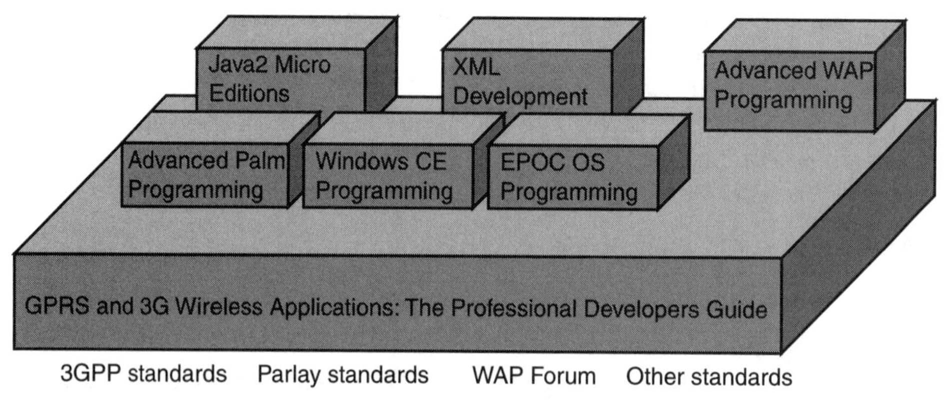

Figure I.1 This book and how it relates to others.

The CD-ROM that accompanies this book contains a number of valuable tools and examples for those who are eager to get started, and by using this CD-ROM, you can get started right away.

What Is on the Web Site and CD-ROM?

The Web site www.wirelessdevnet.com/GPRSand3Gapps contains links to the sites mentioned below and to the latest versions of the tools on the CD-ROM. In addition, there are many tools and SDKs that either had licensing issues attached to them or simply were not available at the time of printing.

The CD-ROM contains a number of useful tools that will get you started. The WAP development kits enable you to quickly get started with developing WAP applications and also contain valuable guides and advice. To complement those features, there are a number of emulators of current mobile devices that are useful for testing the application.

Due to the current state of standardization, the attached Java SDKs are likely to be replaced by updated versions by the time you read this book. The SDKs are still useful and show a bit of what the future will look like. The 3GPP standards (at www.3gpp.org) can be a bit difficult to read for beginners, although they are freely available on the Web. The page www.3gpp.org/3G_Specs/3G_Specs.htm explains the structure and links to the relevant document databases. On the *File Transfer Protocol* (FTP) site, there are directories for the different workgroups and the specifications are grouped by Releases (e.g., ftp://ftp.3gpp. org/Specs/2000-12/). The status file (e.g., ftp://ftp.3gpp.org/Specs/ 2000-12/status_2000-12.zip) contains the document numbers for the standards, and you can use it to locate the correct documents on the FTP site.

cdma2000 is standardized in 3GPP2, and you can find the technical information directly below the main Web page.

The WAP Forum Web site, www.wapforum.org, contains all of the WAP standardization documentation. Most of the information is under the "Technical Information" link.

The Bluetooth specifications are available at www.bluetooth.com. There are two documents for Version 1.0: the main specification and the profiles specification. Note that these files are big, and choosing "Print target" in your browser might make you unpopular among people who are using the same printer.

Parlay is an organization that drives the standardization of open APIs, as we will mention in Chapter 9, "Application Architectures." You can find the Parlay specifications at www.parlay.org.

What Now?

Those who feel unfamiliar with the wireless terminology or who just want to know how I define the different concepts should move on to Chapter 1. Others can dive directly into Chapter 2 and start learning about the mobile evolution and mobile systems.

I hope that many of you will find this book not only useful but also inspiring, and I hope that this information will spur you on to create great applications. The wireless industry needs your contribution to the success of the mobile Internet.

PART ONE

The Mobile Networks

CHAPTER 1

Basic Concepts

For those who enter the world of the mobile Internet for the first time, coming fresh into the business or having some previous experience from the Internet or computer software worlds, it might appear to be a very strange place. The terminology is a mixture of legacies from times that are as distant as the early days of telephony and early twenty-first century wireless nomenclature. The ultimate wireless application developer does not have to master every part of the mobile system and its history, but some basic knowledge is required in order to understand the market, the technology, and most of all the people. Many of the mobile operators that you will encounter as you take your upcoming successes to the market will be deeply rooted in the history of mobile/cellular systems, and you must understand their thinking. This chapter will briefly describe the basics of the technologies that are involved and that are needed to understand the following descriptions. In addition, we will go through some of the concepts that we will use and try to define concepts such as applications and services.

How a Mobile Phone System Works

How much does an applications developer have to know about wireless systems? Understanding how the networks and handsets will affect the performance of the application is crucial, but it is also important to have a basic understanding of the components of the system. In the same way, most *personal computer* (PC) and Internet programmers have a basic understanding (and often more than that) of the hardware and network with which they are working.

Although it might seem like very basic knowledge, even the structure of a mobile system seems unknown to most people. (We will use mobile systems throughout the book to describe cellular, land-borne communications systems.) This lack of knowledge is nothing to be ashamed of; however, most people just would not admit their lack of understanding. I have also heard several questions concerning how the mobile systems talk to the satellites. (Satellites *can* be used, but this situation rarely happens.) In a world where technology is moving so fast, it is of course not realistic to demand that everyone know these things. Rather, a basic understanding makes it easier to follow the reasoning of operators and handset and infrastructure vendors.

Architecture

Figure 1.1 shows a sample mobile system.

This system shows a mobile system with three *Base Transceiver Stations* (BTS), one *Base Station Controller* (BSC), and one *Mobile Switching Center* (MSC). This figure also shows three *mobile stations* (MSs). In a typical network that covers a European country or a U.S. state, there are several thousands of BTSs. The BTSs are commonly called base stations, and sometimes the acronym RBS, Radio Base Station, is used.

Infrastructure vendors such as Ericsson, Nokia, and Lucent develop the mobile system, and a mobile operator buys the system in order to sell the service and airtime to subscribers. The operators usually buy handsets at the same time.

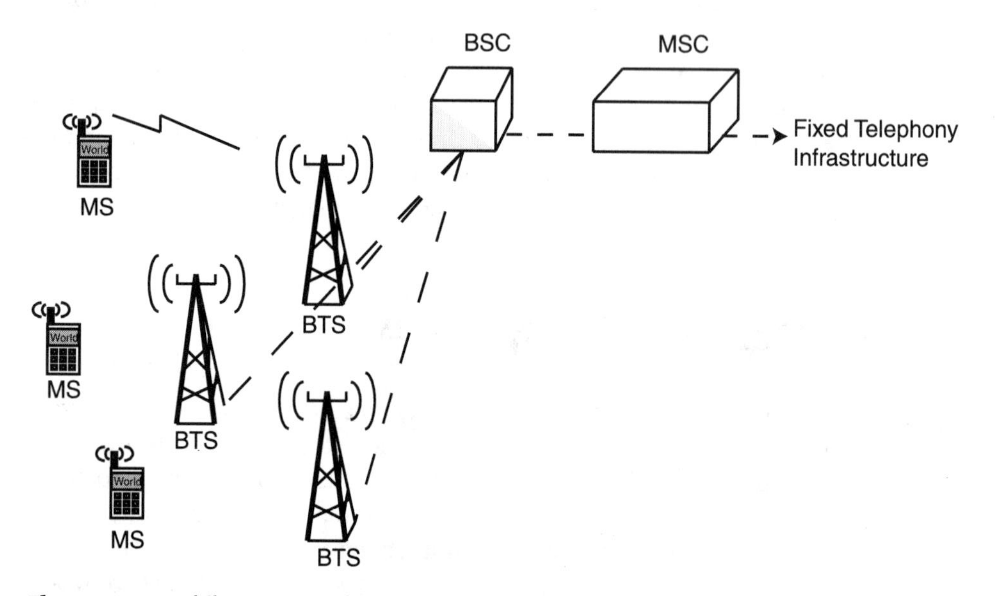

Figure 1.1 Mobile system architecture.

(The majority of handsets are sold this way.) In other words, you never buy a mobile phone subscription from Lucent; rather, you buy one from AT&T, Vodaphone, or whatever operator that serves your area.

Next, we will describe the individual parts of the mobile system in more detail.

The Handset and the Mobile Station

The handset is probably the most well-known piece of equipment, because this is the part we use to make phone calls (and to access data services). When we talk about advanced services, the handset is commonly called an MS, which consists of *terminal equipment* (TE) and a *mobile terminal* (MT). The TE is the device that hosts the applications and the user interaction, while the MT is the part that connects to the network. In Figure 1.2, we show an example configuration where the two parts are physically separated.

Other configurations combine these two parts into one physical, multipurpose device. For further discussions concerning different handset configurations, please see Chapter 10, "Mobile Internet Devices." In some systems, such as *Global System for Mobile* (GSM) communication, *General Packet Radio Services* (GPRS), *Enhanced Datarates for GSM and TDMA Evolution* (EDGE), and *Wideband Code Division Multiple Access* (WCDMA), the subscriber data is stored separately on a *Subscriber Identity Card* (SIM). This feature enables a user to change SIM cards when leaving work in order to convert his or her

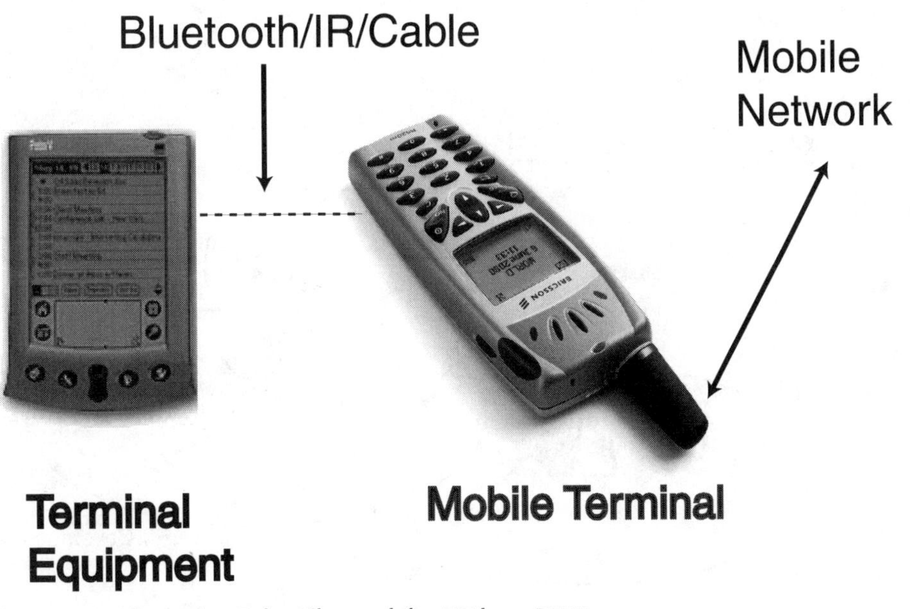

Figure 1.2 The TE is a Palm Pilot, and the MT is an R520.

phone into a private phone that has a private number. The SIM card can also host additional services through the use of SIM Toolkit technology. SIM Toolkit is beyond the scope of this book, however, so we will not describe this concept further.

The Base Station Subsystem

Although the architecture varies a bit between different systems, there is always an antenna that receives signals from the handsets and transports it to the mobile systems. The antennas can be found at various high-level places in order to obtain the best possible coverage. Connected to each antenna is usually a *base station* that processes the call setups and routes the calls to the network. In Figure 1.3 and throughout this book the base station is depicted as an antenna tower—although the core of the functionality lies in a small shed that is usually located at the bottom of the tower.

A *cell* is the basic geographical unit of a cellular system and is defined as the area of radio coverage that one base station antenna system provides. Each cell is assigned a unique number called a *Cell Global Identity* (CGI). The coverage area of a mobile system consists of a huge number of these cells, hence the words *cellular system* and *cellular phones*.

One cell sometimes sends information in all directions from the base station, and sometimes there are three sectors surrounding the antenna. The first configuration is common in rural areas, where it is crucial to obtain as high coverage as possible. The latter configuration, on the other hand, is especially suited

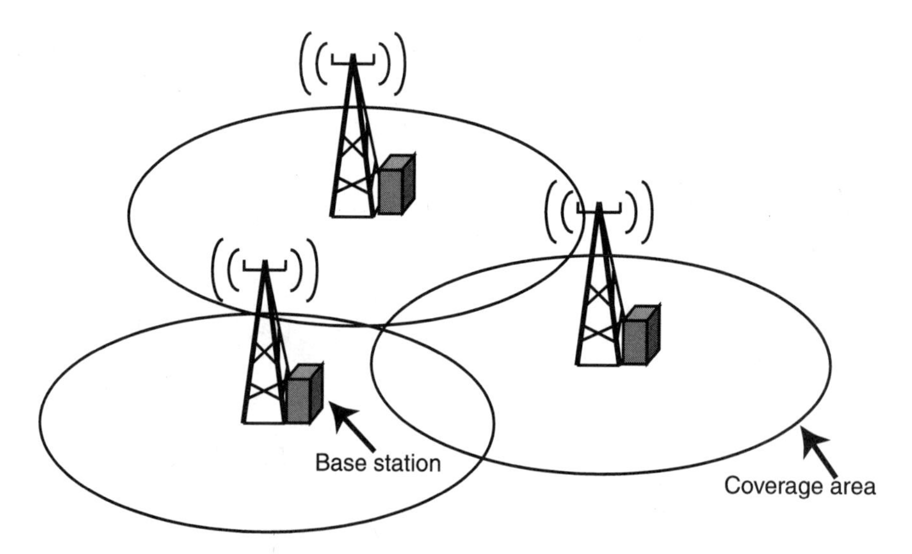

Figure 1.3 Cells and antennas.

for high-traffic areas, and the cells can be directed in clever ways in order to cope with the high traffic. One common example is a stadium, where the load on the network can be incredibly high at times. In these cases, one cell is usually aimed directly at that spot so that it does not deal with any other traffic. So, a base station has an antenna that enables an air interface connection with the MS. When setting up a call, there are commonly some resources (transceiver, power, and so on) allocated to the user in question. One major difference between *second-generation* (2G) and *third-generation* (3G) systems is that the allocation in the base station is much more flexible in 3G. In 2G, there is commonly one kind of resource that is dedicated to a certain kind of service, and this kind of limitation would make a multiservice 3G system very inefficient.

A number of base stations are then connected to a controller (a BSC) for GSM and to a *Radio Network Controller* (RNC) for WCDMA. Much of the intelligence of the mobile system exists here. The BSC/RNC manages all advanced radio-related functions, handover (going from one cell to another), radio channel assignments, *Quality of Service* (QoS), and the collection of cell configuration data. Advanced load balancing and admission control functionality also exists in the BSC/RNC. The controllers and the base stations together are called the base station subsystems.

The Core Network

The core network has traditionally been equipped with switches and subscriber-handling functionalities. These features include subscriber handling, authentication, security, and system maintenance. As more and more advanced services are introduced, the core network becomes more and more of a data network in which circuit-switched and packet-switched services share the same network. As we explain the GPRS architecture in Chapter 3, "GPRS—Wireless Packet Data", this migration will become more obvious. The main task of the traditional core network is to route traffic that enters a mobile network from other networks to the right base station and to route calls from an MS within the system to the right destination network, as shown in Figure 1.4.

The destination network for data services might be another mobile network, a land-line phone network, or the Internet. The advent of advanced data services changes this situation, however, and creates a need for items such as SMS centers, WAP gateways, and so on. We describe these concepts in more detail in Chapter 9, "Application Architectures."

Other Networks

After our call is routed from the MS via the base station, the BSC, and the core network, it now finds the right destination network. The core network switches determine whether the call should be sent to a land-line phone network, to

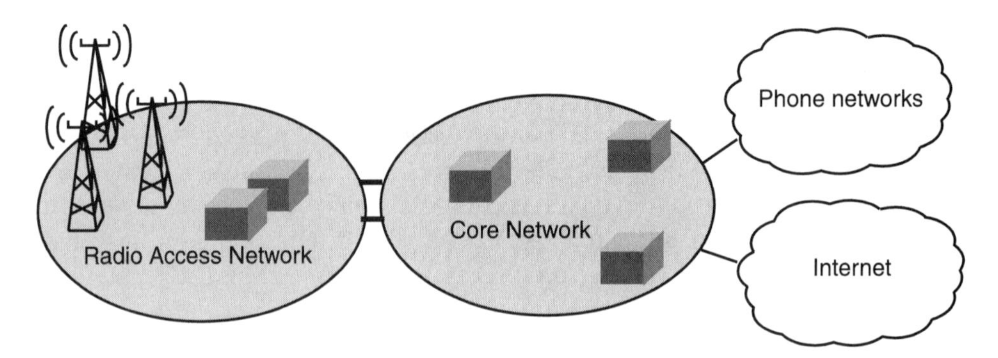

Figure 1.4 The core network transports voice and data to and from the radio access network.

another mobile phone network, or to a different destination. If the destination network is a mobile system, this route is repeated in reverse order. At the base station, the MS is paged with a signal that tells it that someone wants to reach it. You will sometimes notice the paging traffic if your mobile phone is close to a radio when someone calls you. The receiving user's phone rings, and the call can be set up.

As you can see, there are no satellites involved in a regular call with a mobile phone (only in very special cases does this situation occur). The phones do not talk directly with each other; rather, they communicate via networks. The base stations do not send the calls directly to each other; instead, they communicate via a network that most of the time is buried in the ground. Hopefully now you feel confident about how a mobile system works, so we can start getting down to business. As we look at specific systems (GPRS and 3G), you will become more confident by seeing this theory through the perspective of real solutions.

Now, let's agree on some terminology and learn how to place the different technologies with which we are working.

Concepts and Terminology

The terminology that we use in this book is both the widely accepted telecommunications jargon and some of the concepts that are specific to the emerging mobile Internet industry. The latter often lack clear definitions, and different sources use them in various ways. This section aims to remove those ambiguities and to create a set of concepts that we can use consistently throughout this book.

Separating Users from Each Other

In a mobile system, the different users need to use different channels in order to avoid colliding traffic. The three most common ways to achieve this goal are via frequency division, time division, and code division.

Frequency Division Multiple Access (FDMA) gives each user a different frequency. The first analog systems (called 1G, or first generation) commonly used FDMA.

Time Division Multiple Access (TDMA) separates the users in time by assigning different time slots for each channel. Each channel is called a time slot because it allocates a certain time interval during each radio frame. In GSM, there are eight time slots in each frame, giving each user the opportunity to send every eighth time slot (see Figure 1.5). The mobile systems GSM and TDMA (IS-136) use TDMA to separate users.

Code Division Multiple Access is used by the majority of the 3G systems, as well as cdmaOne. In CDMA, different users are separated by different codes. CDMA requires very good power control algorithms, or else only the loudest users would be heard.

Separating Sending and Receiving Traffic

In telecommunications, the words *uplink* and *downlink* are often used to describe outgoing and incoming traffic for the handset (respectively). Figure 1.6 illustrates these two concepts.

Now that we have seen how different users are separated, we need to know how uplink and downlink traffic for one user is separated. The choice of duplex method determines this decision.

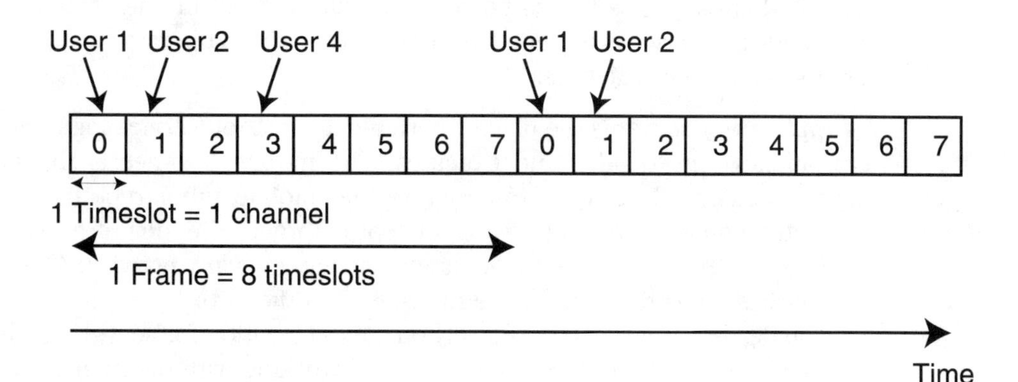

Figure 1.5 GSM time slots.

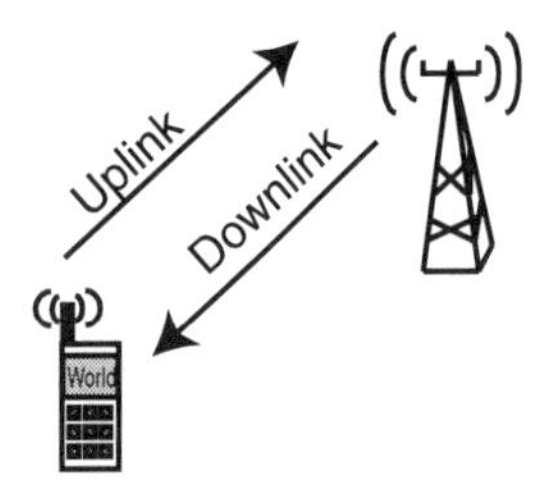

Figure 1.6 Uplink and downlink.

Time Division Duplex (TDD) separates the uplink and downlink channels in time. This is used by Bluetooth, for instance. *Frequency Division Duplex* (FDD) allocates different frequencies for the uplink and downlink channels. WCDMA FDD is an example of how different frequencies are used for sending and receiving.

Defining Concepts

Some of the concepts that we use throughout this book are defined here and are mostly derived from commonly accepted terminology.

An *application* performs a task for an end user (a consumer). This book focuses on applications that are aimed for wireless networks and devices. The application can either be software on the server and/or on the handset or content that is formatted by using scripts and markup languages. Examples of wireless applications include wireless puzzle games, chat applications, electronic calendars, mobile cinema ticket purchases, and so on. We sometimes use the term *service* to describe an application and its surrounding features. As an example, a movie-ticketing application can be used to offer a movie-booking service for mobile users. These two words are often used synonymously, and keeping them separate in this book is difficult, as well. In one sentence, end users consume services provided by applications, which operate on application servers and/or client devices.

An *application server* is the place where the application/service logic for end-user applications resides and executes. The mobile Internet is the merge between existing Internet infrastructure and mobile infrastructure with all associated content. This very wide concept embraces the idea of accessing a large network of information from any device and via any network. The fixed Internet describes the non-mobile Internet. (In other words, applications that desktop PC users and others who rely on a fixed location for accessing the service can access.) In short, we all got used to working with the Internet during the 1990s. A service network is an *Internet Protocol* (IP)-based network that creates a scalable, robust, and secure applications architecture. This network normally consists of application servers, service enablers, and other servers that enhance the applications. The key feature is that both mobile and fixed users can access it from a variety of devices, such as telephones, *personal digital assistants* (PDAs), and desktop PCs.

Summary

The mobile network consists of a radio access network and a core network, and the handset is connected via an air link to a base station. The coverage area of a base station is known as a cell, and a mobile network consists of many base stations. FDMA/TDMA and CDMA are all methods for separating different users from each other, while TDD and FDD separate uplink and downlink traffic. We summarize the definition of applications and services for this book as follows: End users consume services that applications provide, and these applications operate on application servers and/or client devices.

CHAPTER 2

The Mobile Evolution

Many people talk about the mobile revolution, where half a billion people are buying cellular phones each year compared to a few thousand only 10 years ago, and new features are implemented at an ever-increasing speed. Now, we see the start of a similar growth of the mobile Internet, where Japan is adding hundreds of thousands of subscribers every month, and the rest of the world is producing small startups in the area at an amazing pace. Although there are undoubtedly many things happening at an incredible speed, we will still look at this revolution more like an evolution. We will see how the emerging systems are developed as enhancements to existing systems, rather than as radically new systems.

This chapter will briefly describe the history of the mobile Internet—mainly, its cellular system roots as its fixed Internet counterpart in Chapter 6, "Unwiring the Internet." This chapter aims to create a better understanding of why we have several *second-generation* (2G) and *third-generation* (3G) systems and how the different operators are likely to migrate to 3G.

Mobile Phone History

Although commercial mobile telephone networks existed as early as the 1940s, many consider the analog networks of the late 1970s in the United States (in Europe, it was the early 1980s) to be the *first-generation* (1G) wireless networks. These networks were designed similarly to fixed-side networks, where an analog image of the sound was transmitted over the air and through the networks. The receiver and transmitter were tuned to the same frequency, and the voice that was transmitted was varied within a small band to create a pattern that the receiver could reconstruct, amplify, and send to a speaker. Although this technology was truly a revolution in the area of mobility, these systems had some serious shortcomings. Users who wanted to travel became disconnected as they moved out of their coverage area, and therefore they had to reconnect. The handover feature, which makes it possible for a mobile phone to seamlessly switch the antenna from which it receives and sends, was not available—and this lack of technology seriously limited mobility. Another problem was the lack of efficiency, because very few callers could fit into the available spectrum (in this book, the word *spectrum* describes a frequency range that the mobile system uses). Analog systems are generally not fit for optimizations such as compression and coding. The components that were used were also big and expensive; consequently, the handsets looked more like bricks than telephones.

Despite these challenges, the analog systems were successful in the United States (measured by the standard of those days), and consumers could use a common handset across the North American continent (as long as coverage existed). The *Advanced Mobile Phone Service* (AMPS) was first introduced (trials) in New Jersey and Chicago in 1978, and the interest in the technology spread to various parts of the world. The introduction in North America awaited some regulatory issues that the *Federal Communication Commission* (FCC) had to solve, and countries such as Saudi Arabia, Japan, and Mexico got things up and running before the United States did. After the first commercial launch (Ameritech, 1983), the takeup began all over the United States. As the systems grew, operators saw the added complexity and system load—which resulted in the development of a common standard for the core networks: TIA-IS-41 (often just called IS-41). The core network is the back-end infrastructure that transports a voice call to and from a mobile user's radio network to another mobile/fixed user. As the wireless industry started to discuss how it could introduce a digital system, a major concern was to keep legacy support for the analog systems while drastically increasing the capacity and introducing digital transmission.

In Europe, the countries struggled with no fewer than nine competing analog standards during the 1980s, such as Nordic Mobile Telephony (NMT), Total Access Communications System (TACS), and so on. Pan-European roaming was nothing more than a distant dream at this point, and capacity became an increasingly difficult issue. Europeans therefore saw the need for a completely new system—a system that could accommodate both the increasing subscriber base as well as more advanced features and a standardized solution across the continent. Because of the shortcomings and incompatibility issues with analog systems, they decided to institute a completely new digital solution. The new standard, *Groupe Spéciale Mobile* (GSM), was built as a wireless counterpart of the land-line *Integrated Services Digital Network* (ISDN) system. Although

GSM initially stood for Groupe Spéciale Mobile, named after the study group that created it, the acronym later changed to stand for Global System for Mobile communications. This occurrence would not be the last time in the history of mobile systems that an acronym would change, as we will see later. Twenty-six European national phone companies standardized the system, and the working process set the standard for a way of working that has proven successful many times. The countries and the individual companies realized the power of a cross-border standard and the kind of money and energy that can be wasted when competing for world domination on your own.

The results of this and other projects related to "going digital" led to four major 2G wireless systems. *Digital AMPS* (D-AMPS) was a digital add-on to AMPS (which we now call TDMA). With D-AMPS, the handset can switch between analog and digital operation. IS-95, a CDMA-based solution that Qualcomm introduced in the mid 1990s, picked up toward the end of the century. IS-95 is now more commonly called cdmaOne. In Europe and Asia, GSM quickly became the dominant standard with a high degree of extra services, such as the popular *Short Message Service* (SMS). In Japan, *Personal Digital Cellular* (PDC) became the number one system. However, this system put Japan in an awkward situation, with an old system that was incompatible with all of the others. This situation triggered the Japanese operators to start an aggressive pursuit of new technology and standards. In the late 1990s, cdmaOne began gaining ground in the Japanese market, increasing the pressure even more on existing PDC operators. Table 2.1 and Figure 2.1 show the distribution between the different systems as of fall 2000.

With the advent of digital systems, the sound of the speaker's voice was sampled and filtered through various advance speech models, which basically imitate a human ear. The resulting 1s and 0s were sent over the wireless network to the receiving party. A digital mobile user who received a call would hear the reconstructed voice, created by the digital signals passing filters that imitate the human speech system (vocal cords and so on). The digitalization made it possible to squeeze more subscribers into the same radio spectrum, thus increasing efficiency. In addition, the advances in digital chip technology facilitated the development of small and light handsets that boasted an ever-increasing degree

Table 2.1 Number of 2G Subscribers as of August 2000 (per System)

SYSTEM	NUMBER OF SUBSCRIBERS IN MILLIONS
GSM	362
cdmaOne	72
PDC	48.8
TDMA (IS-136)	54.3

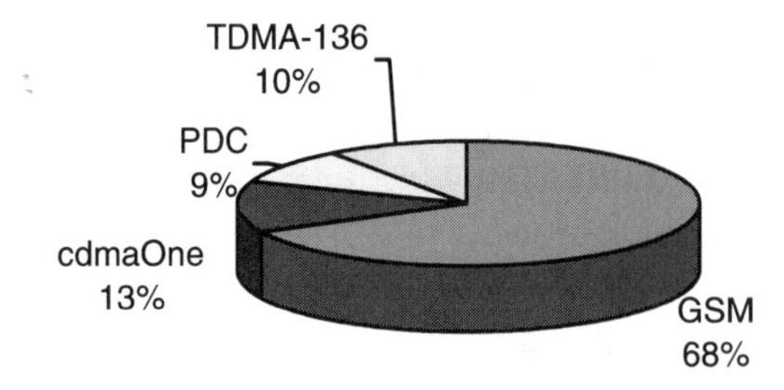

Figure 2.1 Number of subscribers of 2G systems as of fall 2000.
Source: EMC World Cellular

of features. These features included voice mail, call waiting, and advanced supplementary services such as SMS. SMS uses the control channels of GSM, TDMA, and cdmaOne systems to transmit messages up to 160 characters long. In the late 1990s, the GSM operators saw an amazing increase in SMS usage. In late 2000, there were almost 15 billion SMS messages sent every month and one billion in Germany alone. Even more noteworthy was the fact that a new segment of the market—teenagers—had taken the lead in using these advanced services. The SMS messages are perfect for communicating in environments where it is hard to hear each other, such as in nightclubs. They also are a quick way of notifying others without entering a long voice call.

The major driving force behind the 3G wireless systems was once again the need for capacity and global roaming, but this time, the motivation was also higher bit rates and a higher *Quality of Service* (QoS). The work on 3G concepts started in the early 1990s, and in parallel, the Internet wave started to catch on. Therefore, the initial vision was to create a global wireless system with high speed and quality as complements that would fit the need of a mobile Internet. In the struggle to achieve global roaming, the legacy systems were again a major obstacle. With several hundred million mobile users, no one wanted to abandon those subscribers. In the United States, there was another problem with the frequency allocations. Previous auctions of the spectrum that paved the way for the digital 2G systems in the country had effectively blocked the implementation in the 2GHz band (the PCS frequency band is 1900MHz, or 1.9GHz). Japanese and European operators and infrastructure vendors had planned to use this band for 3G. An intensive battle ensued as more and more players started to realize how much was at stake. After many rounds of discussion, they decided that there would be three main branches of the 3G standard and that a convergence effort would begin. The three standards are WCDMA, CDMA2000, and *Enhanced Datarates for Global Evolution* (EDGE), where

WCDMA actually has two different modes (FDD and TDD). We will describe the 3G standards in the next chapter.

Mobile Systems Now and in the Future

Now that we have seen how a mobile system works, it is time to examine the new systems and determine how to migrate to them. The key here is that the old systems are not going to be thrown out just because the new ones are introduced. The mobile operator (sometimes called the carrier) is the one that sells you a subscription (AT&T, Vodaphone, Telia, and so on) and that has invested massive amounts of money into this infrastructure. Carriers do not want to waste everything. Also, the infrastructure was built to cover as much geographical area as possible in order to provide users with good coverage. Building completely new systems means that coverage would need to be rebuilt from scratch. So, a smooth evolution is preferred as opposed to installing completely new systems. Now that there are four different 2G systems, and the availability of frequencies is different, there must be different ways for operators to migrate to 3G systems. Figure 2.2 illustrates the different migration paths. Note that there are creative ways of getting to the desired 3G standard that this picture does not capture. For example, AT&T started with a TDMA network but then decided to have GPRS, EDGE, and WCDMA and therefore installed a GSM network as well in order to get there.

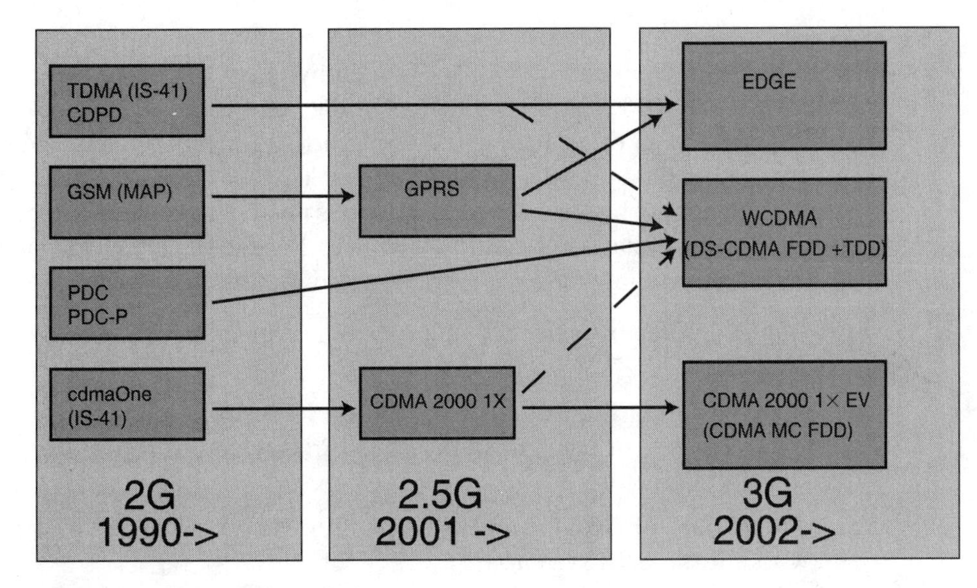

Figure 2.2 The mobile evolution.

Now, let's look briefly at these different evolution paths and where they will be used.

GSM Evolution

GSM is by far the biggest 2G system, with more than 400 million subscribers (by the end of 2000) and adding 10 million more each month. With its pan-European coverage and systems also installed in Asia, Australia, and North America, GSM is now truly a global system. In 2000, GSM also started to gain momentum in South America. With that in mind, being compatible with GSM from day one was a prerequisite for any new system that would add functionality to GSM. As with other 2G systems, GSM handles voice efficiently, but the support for data and Internet applications is limited. A data connection is established in just the same way as a regular voice call: The user dials in and a circuit-switched connection continues during the entire session. If the user disconnects and wants to reconnect, the dial-in sequence has to be repeated. This issue, coupled with the limitation that users are billed for the time that they are connected, creates a need for packet data for GSM.

With *High-Speed, Circuit-Switched Data* (HSCSD), the same circuit-switched technology is used as regular GSM, but multiple timeslots can be used for one connection. In other words, one user can achieve up to 57.6Kbps of data speed. The first HSCSD systems appeared in 2000, and the first batch of terminals was in the form of PC cards. This format enables users who are already frequently checking their mail and using the Internet/intranet on the go to achieve higher speeds.

Packet data is introduced into GSM systems by using *General Packet Radio Services* (GPRS). GPRS is an overlay technology that is added on top of existing GSM systems. In other words, the GSM part still handles voice, and handsets are capable of supporting both functions. The GPRS upgrade is easy and cost effective for operators, as only a few nodes need to be added. We deal with GPRS in detail in Chapter 3, "GPRS—Wireless Packet Data," but for now we will stay with the following three key features of GPRS:

Always online. Removes the dial-up process, making applications only one click away.

An upgrade to existing networks (GSM and TDMA). Operators do not have to replace their equipment; rather, GPRS is added on top of the existing infrastructure.

An integral part of EDGE and WCDMA. GPRS is the packet data core network for these 3G systems.

While GPRS is an obvious migration step for GSM operators, the next step requires further evaluation. The two main tracks to pursue are EDGE and WCDMA.

EDGE is a cost-efficient way of migrating to full-blown 3G services. EDGE does not change much of the core network, however, which still uses GPRS/GSM. Rather, it concentrates on improving the capacity and efficiency over the air interface by introducing a more advanced coding scheme where every time slot can transport more data. In addition, EDGE adapts this coding to the current conditions, which means that the speed will be higher when the radio reception is good.

A key feature of EDGE is that commonly no additional spectrum is necessary, and EDGE boosts the capacity and bit rates of existing GSM/GPRS as well as TDMA systems. We describe the EDGE system in more detail in Chapter 4, "3G Wireless Systems."

WCDMA is another migration path that you can use with or without EDGE. Some operators who have acquired WCDMA licenses will still invest in EDGE to gain the following competitive advantages:

- Faster time to market with 3G services by offering EDGE during late 2001. This offering will make it possible to catch early adopters and then keep them when introducing WCDMA.
- Once WCDMA is introduced, it will initially have limited coverage. The fallback solution in rural areas will be GPRS or EDGE, where EDGE provides significantly higher bit rates and capacity.

WCDMA, or UMTS (Universal Mobile Telecommunications System),which it is called in Europe (UMTS is also commonly used to describe the 2GHz frequency band needed), is an upgrade to GPRS/EDGE on the core network side, and it also introduces a completely new radio interface. The new CDMA radio interface uses codes to separate users instead of the time slots that TDMA systems such as GSM/GPRS use. While the interface is completely new and different from GSM/GPRS, even the first batch of WCDMA handsets on the market are predicted to be compatible with GSM/GPRS. This compatibility is crucial, because the new radio interface means that coverage has to be built from scratch once again. It is foreseen that even many years after the introduction of the first WCDMA system, rural areas will be limited to EDGE or GPRS coverage. This detail is an important one to consider when designing applications. How do I handle a graceful degradation of service when the 400Kbps of WCDMA is replaced by 20Kbps of GPRS? Also, the nature of CDMA systems makes the bit rate highly dependent on the distance to the base station. In other words, even within a WCDMA system you are likely to achieve lower bit rates as you move away from the antenna. The RNC is already remedying part of this phenomenon. The RNC can distribute the capacity in a cell according to the QoS information that is associated with the subscribers. Thus, a user who is paying a premium can keep a constant bit rate as long as his handset can deliver the power that is needed to reach the antenna. We describe WCDMA in more detail in Chapter 4, "3G Wireless Systems."

TDMA (IS-136) Evolution

In TDMA, packet data is already introduced in the form of *Cellular Digital Packet Data* (CDPD). It all started in 1992, as a group of U.S.-based operators with AMPS-standard wireless networks formed a consortium to steer the introduction of data services. CDPD technology enables D-AMPS/AMPS carriers to offer both voice and mobile Internet services, leveraging the same network infrastructure and channels.

CDPD is a cost-efficient add-on for TDMA operators, because only a small functional upgrade of the base stations is necessary. Being a packet data network, CDPD can run *Internet Protocol* (IP) applications and can act as an extension of the Internet, where users can be constantly connected (similar to GPRS). As a consequence, each CDPD mobile is assigned an IP address, and all user packets in a CDPD backbone network consist of IP packets. In the early days of CDPD, this system was mainly used for vertical applications, and there were no devices or services available for the network. As WAP emerged and the increasing hype for the mobile Internet grew in the late 1990s, the operators started to look at ways of bringing CDPD to the mass market. In May 2000, AT&T introduced its PocketNet service, running WAP-like (HDML) services over CDPD. Consumer-oriented devices appeared at the same time, and suddenly CDPD was a consumer network technology. An interesting aspect of this relaunch of CDPD as a consumer-oriented network was the consumer's choice of handsets. Most people started with a voice-oriented handset in the low-end price range and saw the mobile Internet features as a bonus. While developers might have wanted a bigger spread of the high-end handsets with big displays, the consumers were not ready to spend the extra money. This situation teaches us an important lesson about people's abilities to adopt new technologies. Even if there is something new that is useful and compelling, the users need to be slowly migrated into this technology. In this case, the move from a voice-oriented phone to a high-end *personal digital assistant* (PDA)-like device was too big, and users were more likely to choose this kind of phone as their second or third choice for that technology.

A major concern about CDPD is the lack of upgrade paths toward 3G. To remedy this problem, EDGE was directed toward a version that could run on TDMA networks. A key advantage with this approach is that GSM and TDMA users can roam into each other's networks, taking us closer to the goal of using one handset wherever we go. In order to facilitate this function, two different but still compatible versions of EDGE were necessary: one that uses the TDMA channel structure and one that uses the GSM channels. Existing TDMA operators might choose to go for GSM and its upgrade path as well, depending on spectrum availability. You can find information about the two modes and other details about EDGE in Chapter 4, "3G Wireless Systems."

Today, TDMA is mostly used in North America, by operators such as AT&T, Cellular One, and so on. Telefonica and others serve TDMA in South America.

cdmaOne Evolution

cdmaOne has a similar migration path as GSM, which involves higher circuit-switched bit rates, always-online networks, and higher speeds. You might find it harder to fully understand the cdmaOne migration path, however, because air interface and core network evolutions are clearly separated. See Figure 2.3 for information about cdmaOne evolution.

Figure 2.3 attempts to illustrate how the core network has its own migration, and it can be done independently of the air interface. Generally, it can be said that the full 3G core network of cdma2000, based upon Mobile IP, is more advanced than GPRS and is probably closer to WCDMA phase 2 (introduces an all-IP core network). Some operators might choose to skip the Simple IP step and implement cdma2000 with Mobile IP directly.

The cdmaOne radio interface IS-95A supports voice calls and data rates of up to 14.4Kbps. SMS is also available but has never become the success that it has with GSM. With IS-95B, the overall functionality is improved, but the major step is increased data rates. By combining several 9.6 or 14.4 channels, up to 115.2Kbps can be achieved. As always, this rate is lower in reality and is limited by handsets and the overall capacity of the system. In Japan, services that have bit rates of up to 64Kbps have been launched.

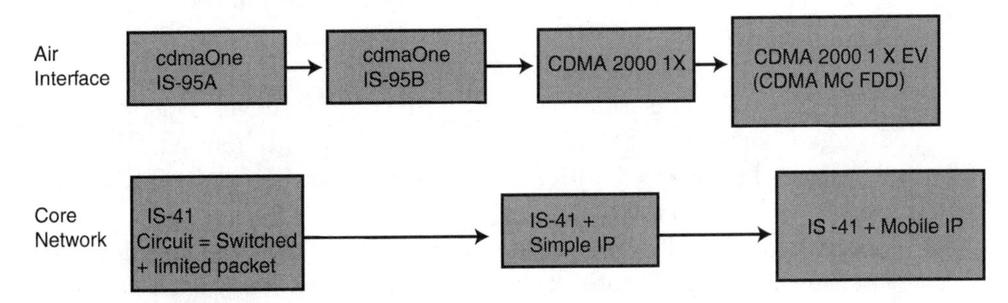

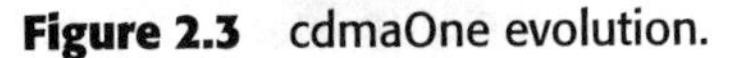
Figure 2.3 cdmaOne evolution.

Most cdmaOne networks are implemented in the 800Mhz and 1900Mhz (PCS) bands and are most widespread in the United States, Korea, and Japan. In the United States, operators include Sprint PCS and Verizon, and cdmaOne has experienced a rapid growth in recent years. When considering the introduction of 3G, the United States has a bit of a problem. The PCS band already occupies the 2GHz frequency band that the *International Telecommunication Union* (ITU) recommends. Therefore, the 3G migration for cdmaOne works in the existing spectrum, rather than requiring a new spectrum to be freed.

With cdmaOne technology, each carrier (channel) is 1.25MHz wide (a GSM carrier is 200kHz and a TDMA carrier is 30kHz). The cdma2000 1X radio interface is backward compatible with IS-95A and IS-95B and therefore uses the same 1.25MHz channels. Through improved modulation, power control, and overall design, cdma2000 1X provides average bit rates of up to 144Kbps (commercial deployments will show the actual values) and also gives the operator more capacity, both for voice and for data. Developers then introduce the cdma2000 3G technology as an overlay to the existing system, where each user has three 1.25MHz channels. The terminology differs, and some call cdma2000 1X a 3G system while some compare it with GPRS and call it "evolved 2G or 2.5G." People sometimes call cdma2000 multicarrier "cdma2000 3X" or "cdma2000 1X evolution."

The core network of cdmaOne is based on the same IS-41 core network as TDMA. The vision of the cdma2000 3G core network is to have a network architecture that is solely based on *Internet Engineering Task Force* (IETF) IP standards with seamless connectivity (called Mobile IP). We will describe Mobile IP in Chapter 4, "3G Wireless Systems."

PDC Evolution

In Japan, NTT DoCoMo was interested early on in introducing high-speed multimedia services to its subscribers. Because Japan was isolated with its PDC system for 2G, NTT and other Japanese companies wanted to make sure that this situation would not happen again. Early on, there was a close cooperation between different companies across the world in order to facilitate more of a global standard for 3G than what had been around for 2G. In the beginning, the Japanese, Association of Radio In Business (ARIB), and European Telecommunications Standards Institute (ETSI), suggestions for the standard diverged somewhat. After a few rounds of negotiations and harmonization, they agreed to proceed with a common standard in a global forum: the *Third-Generation Partnership Project* (3GPP). During this time, NTT DoCoMo introduced its own packet data add-on to PDC, P-PDC, on which I-Mode (described later in this chapter) runs. The main difference between the Japanese migration and the rest of the world's migration is that the step to 3G is much faster.

Japanese operators plan to introduce commercial WCDMA services during spring/summer 2001, which is ahead of Europe and the rest of Asia. The Japanese market was somewhat left out in 2G, and it definitely does not want to repeat that record. The expected strong Japanese market is also likely to fuel the Japanese device manufacturers, and there certainly will be many exciting devices emerging from them.

WAP, Bluetooth, and Other Related Evolutions

So far, we have only talked about the networks and how they will evolve. Other technologies must evolve in parallel, however—technologies that can operate independently of EDGE, cdmaOne, WCDMA, and so on. We commonly call these technologies *bearers*, and they enable the transport of information over the air.

In order to transport information over a bearer, there must be mechanisms for deciding how and where to send the information. Protocols such as IP, TCP, User Datagram Protocol (UDP), and Wireless Session Protocol (WSP) perform some of these tasks, and we describe these protocols in more detail in Chapter 6, "Unwiring the Internet." These protocols also can make sure that the information is received reliably and in order. The *World Wide Web Consortium* (W3C) standardizes most of the Internet protocols, and there is currently work going on to improve protocols such as TCP in order to better handle the wireless environments. WAP is a suite of protocols that will evolve as we get larger displays and more CPUs for devices. You must understand that WAP can evolve independently of the bearers, and WAP can run over SMS and GSM as well as on GPRS, Bluetooth, cdma2000, and on most other networks. We will provide more information about WAP in Chapter 7, "The Wireless Application Protocol (WAP)," and about Bluetooth in Chapter 5, "Bluetooth—Cutting the Cord!"

The Mobile Internet

The mobile Internet—the marriage between today's Internet and the increasing urge for mobility—is about to take off. Independent sources all estimate incomprehensible markets for these new services and applications. Ericsson recently revised its prognosis and now estimates that there will be some 600 million mobile Internet users in 2004 (see Figure 2.4).

Figure 2.4 only illustrates the start of this convergence, and in the long term it will be difficult to distinguish the different parts. Many applications will be accessible across fixed networks as well as mobile ones.

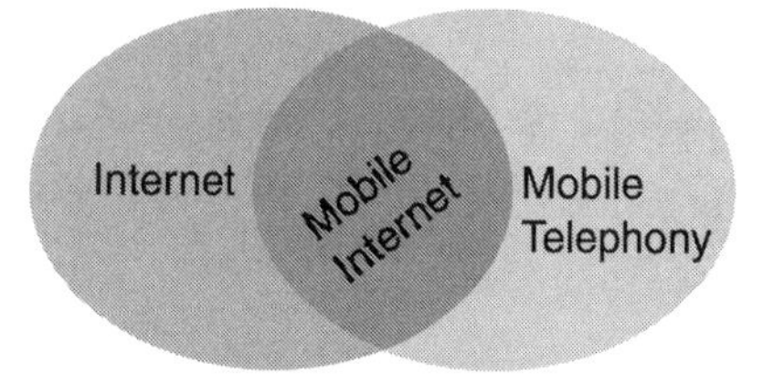

Figure 2.4 Internet + Mobility = Mobile Internet.

Infrastructure and handset manufacturers such as Ericsson, Nokia, Motorola, and so on—and the wireless operators, such as Vodaphone/Airtouch, Sonera, AT&T, and more—have set the groundwork for this paradigm shift. All of them are committed to deploying GPRS and 3G networks. Now, these players feel confident that their parts of this new market will be in place (in other words, technologies such as WAP, Bluetooth, GPRS, and 3G). This situation raises a need to mobilize other players to contribute: the applications (mail, games, chat, and so on) and content developers (Web, WAP, and so on). The mobile Internet is still in its infancy, and most people have still not realized what it will look like and how we will use it. As we will see as we go through this book, the width of possible applications is endless, but there are certain things that make the mobile Internet totally different from its fixed counterpart. By looking into some of the success stories of the mobile Internet so far, we will see what components are crucial.

Mobitex and Palm.net

The specification of the Mobitex technology started at the Swedish Telecom around 1982, and the first Mobitex network entered commercial operation in Sweden in 1986. During 1988, Ericsson became responsible for the further development of the Mobitex infrastructure together with the *Mobitex Operators Association* (MOA) organization. MOA is responsible for the Mobitex Interface Specification, in which the air-protocol and the protocol for the permanently connected terminals are specified.

Mobitex is a network technology that is designed exclusively for two-way, wireless data communication. Mobitex has, since its commercial start in 1986, evolved into a high-quality system of numerous functions and applications. The system uses a hierarchical cellular infrastructure, which you can configure for a wide range of network sizes (giving operators the flexibility for incremental expansion of host connections, coverage area, and message capacity as required). The bit rates of Mobitex are low (8Kbps), but a key advantage is that it has the always-online packet data feature. As with CDPD, Mobitex lacked consumer-oriented applications and terminals for a long time. In 1999, Palm

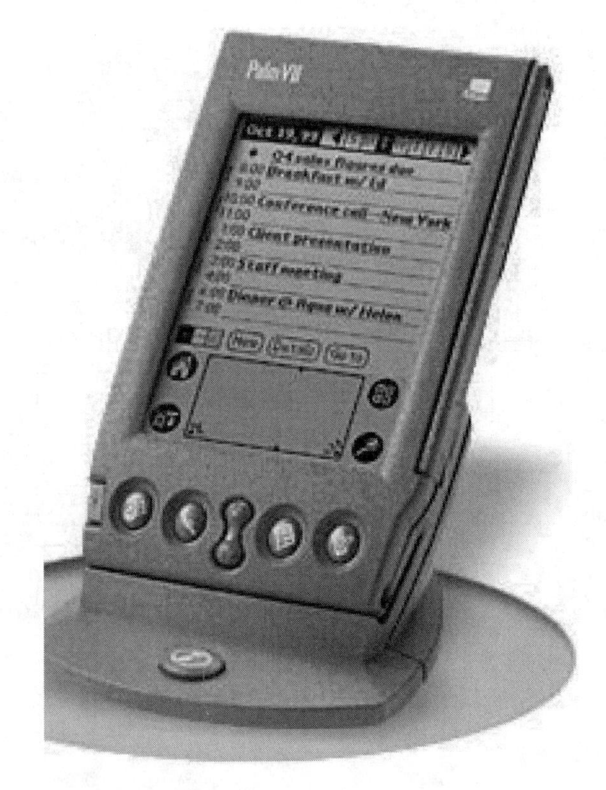

Figure 2.5 The Palm VII.

Computing (at that time a part of 3Com, Inc.) introduced a wirelessly connected device, the Palm VII, with an associated service, Palm.net, that ran on Mobitex networks (see Figure 2.5). With the help of a large set of content partners, this pioneering of using wireless PDAs created a lot of interest.

I-Mode

As NTT DoCoMo was driving the work to produce a 3G standard to deliver multimedia applications to its customers, NTT wanted to start even earlier than the promised launch dates of 3G. Using a packet data extension (enabling Always Online) to its PDC network, a markup language (c-HTML), and lots of content partners, DoCoMo launched I-Mode in March 1999. Figure 2.6 shows the impressive takeup of these services (more than 15 million at the end of 2000).

Users pay a fixed monthly fee and then a price of 300 Yen per packet. The most popular applications are in the entertainment segment, with games, horoscopes, and cartoons topping the charts. Many said that they did not sign up for I-Mode because it would make them more productive, but rather they wanted to have fun. Much debate exists concerning whether these experiences are

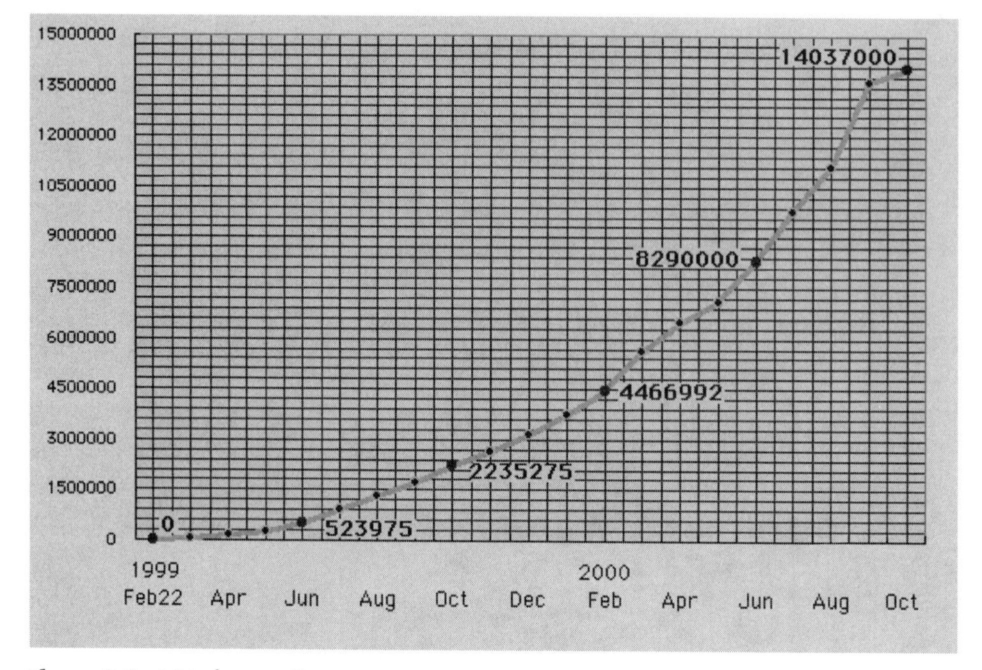

Figure 2.6 I-Mode growth.

directly transferable to the rest of the world. Some argue that the low Internet penetration in Japan has made the Japanese turn to their mobile devices for those services that many others can access from their home computer. Others claim that the enterprise market and business-to-business (B2B) is much more pervasive in, for instance, the United States—implying that entertainment will not necessarily be the most popular application everywhere. It is however important to see I-Mode as more than just a technology and a mark-up language. Rather, it is a concept and a way of doing business by giving the users what they want. In the process, the application developer can easily charge for the services, and this open approach from NTT DoCoMo has been a significant success factor.

As of late 2000, more than a million new I-mode subscribers were added every month. Many wireless evangelists have used this success as an example of the incredible potential of the mobile Internet.

What Makes the Mobile Internet Take Off?

Looking at the examples of I-Mode of Palm.net, we can see some common denominators that created a massive increase in takeup. The three major factors are as follows:

1. **Availability of compelling content and applications.** There has to be something that makes people want the new service, and the funny jokes rather than an increased bit rate will be what drives the growth.
2. **Availability of mass-market devices at reasonable prices.** The content will most likely never be developed unless the content and application developers see some devices that their target users are likely to embrace. The devices must also be incredibly user-friendly, just like the I-Mode button on the Japanese handsets.
3. **Ease of use and hassle-free connections.** The packet data networks that remove the dial-up sequence make the services available at all times (without a tedious initial waiting period).

Another key factor to consider is the time that it takes for users to start accepting a new technology. Never in history have new technologies been introduced to people this quickly. After six months, people started wondering whether WAP was dead just because there were not tens of millions of WAP users. We will have to learn that it takes time and that not everybody is living life in the fast lane.

The third listed item leads us toward taking a closer look at how these packet data systems work and how a developer can use them in the most efficient ways when developing applications.

Summary

The mobile evolution shows us how the systems of today and tomorrow (to a large extent) are designed with legacy systems in mind. Creating a smooth migration for both users and operators is essential in order to succeed. The mobile Internet was introduced as the merge of the fixed Internet and the mobile telephony worlds, where the services and applications are expected to grow together—making it hard to separate them. In order for the mobile Internet to succeed in a market, there need to be handsets, applications, and packet data networks.

CHAPTER 3

GPRS—Wireless Packet Data

We have found that one of the key success factors for the mobile Internet is the access to packet data networks. This chapter focuses on the introduction of packet data to cellular networks. As we described in the previous chapter, the different networks have different ways of getting there, but the end result is similar: users can be constantly connected to the networks without having to pay by the minute. Throughout this chapter, we use *General Packet Radio Services* (GPRS) for *Global System for Mobile communications* (GSM) networks due to its extensive spread, but most of the discussions are relevant for Mobile IP, which is the core network solution for cdma2000 networks. In areas where the difference affects applications developers, we clearly note this information in the text. The first two subchapters give a high-level overview that is sufficient for executives and for others who do not want to dig into too much of the technical details.

The Need for Packet Data

Why do we need packet data, and what is packet data really? As the *Wireless Application Protocol* (WAP) began to spread throughout the world during 2000, some users complained that it was slow, expensive, and cumbersome to use. The fact is that most of the characteristics of WAP over GSM that users complained about were not due to bad WAP performance; instead, these problems are typical for circuit-switched networks. A circuit-switched connection to some site works just like a regular telephone call. You dial in to your *Internet Service Provider* (ISP) and have your 9.6Kbps as long as you are connected (and do not share this capacity with anyone). For certain streaming sound applications, this solution might be a good one (although most streaming applications are not of a constant bit rate). For bursty sessions such as WAP browsing, however, it is inefficient for both the user and for the operator.

Figure 3.1 shows how a radio channel (time slot 2, TS 2) is assigned to the user and displays the data that is transferred. Effectively, the user is paying the same money when sending as much as he or she is when silent. One time slot equals the capacity that is required for a voice call, and there are eight such time slots per transceiver unit in the base station. Circuit-switched data always requires at least one time slot to be allocated during an entire data session, regardless of how much data is actually transmitted.

Similarly, the operator is not utilizing his or her ability to the fullest, because no one else can access the unused channel. This situation is only part of the problem with using circuit-switched data. As we just described, the user also has to establish a new connection when he or she wants to get some information (if not already connected). Say that someone wants to check the weather by using a WAP phone and, when he or she is done, disconnects. If that someone then wants to check the weather in a neighboring area five minutes later, he or she has to establish a new connection. This connection establishment process sometimes takes as much as 20-40 seconds, but you can cut it down to 5-10 seconds by

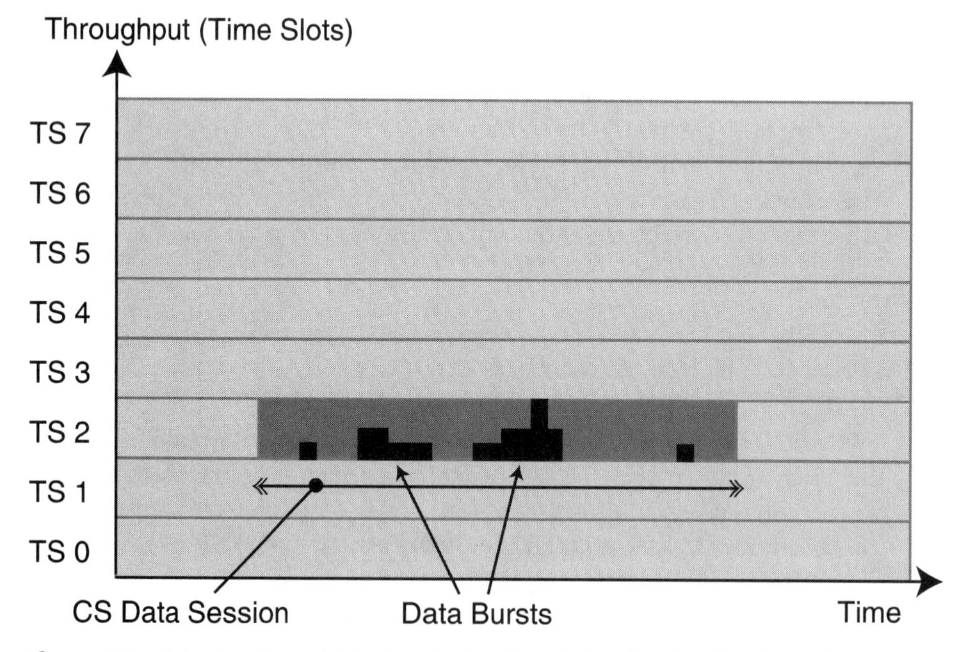

Figure 3.1 An entire time slot is allocated to data, but only a fraction is used.

using special access routers. To summarize, circuit-switched networks are less suitable for data sessions where you do not need a guaranteed bit rate and where the amount of information that is sent and received varies greatly. The cost is high for both the user and the operator.

Introducing packet data on a network not only solves these problems, but also enables users to share the radio resources (just like traffic is handled over the fixed Internet, where several users share the same connection in order to maximize efficiency). This way, you are not putting any load on the network when you are not sending or receiving packets. With GPRS, we take this thinking one step further. With packet data, users not only share the capacity with each other, but also share it with circuit-switched voice and other data users. We illustrate this example in Figure 3.2, where two GPRS packet data users share the first two time slots (they could have shared one as well). Two more users who have the same usage characteristics could probably have shared the same two time slots without any perceived degraded speed. In Figure 3.2, five circuit-switched channels are also allocated during a period. They utilize time slots three through seven and are not affected by the GPRS sessions.

This feature is also beneficial for the operator, who now can accommodate more users within the same network. In addition, this feature makes it possible to utilize parts of the network that you could not use before. A little known fact is that a circuit-switched network that is fully loaded, where some users get blocked calls (you cannot make a call because there are no resources), can still have as much as 40 percent of unused capacity. This situation results because

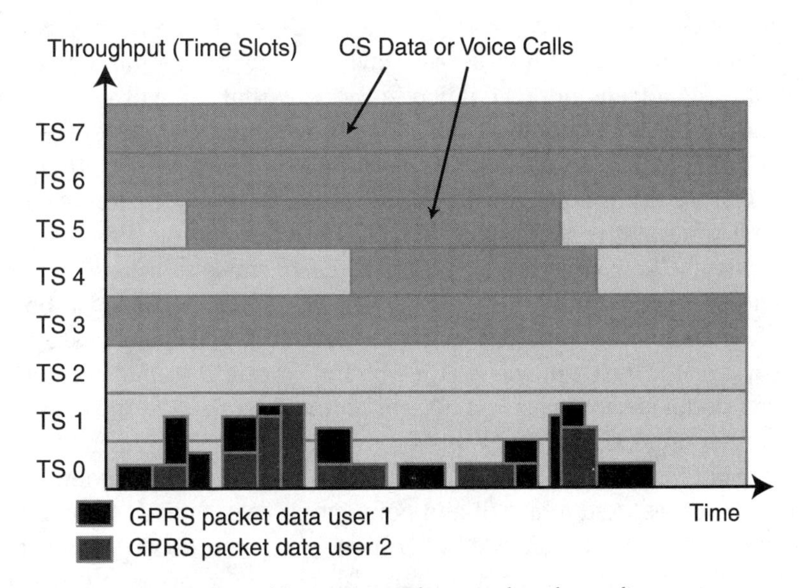

Figure 3.2 GPRS enables users to share packet channels.

of the gaps that are created between the disconnection of a channel and someone else who is connecting. A good analogy is to think of a circuit-switched network as a large box of rocks that can never be 100 percent full due to the irregular shapes of the rocks. Filling the same box with rocks and sand (or sand alone) makes it possible to fill up the box more than before. With GPRS, however, the number of users goes up, and added interference is the result. This situation makes it hard to get 100 percent utilization, but it is still a significant improvement. Now, let's look at this GPRS sand and how it works.

Key Features of GPRS and Packet Data Networks

The following three key features describe wireless packet data, GPRS, in a nutshell (packet data for cdma2000 has similar features):

- The always online feature
- An upgrade to existing systems
- An integral part of future 3G systems

In detail, the following paragraphs describe what these features mean for the user and companies that are involved.

The Always Online Feature

As we described previously, GPRS enables the user to be always connected and always online without necessarily having to pay by the minute. This functionality means a tremendous change in the way that we will use the cellular phone and probably the home PC once this functionality is introduced there as well. You can use this feature with a *Digital Subscriber Line* (DSL), for example. This feature is not unique to GPRS; however, the introduction of GPRS will be the first time that most mobile Internet networks obtain this feature. This functionality is then kept for all 3G systems because it is one of the most important features of any wireless network of the future. You will be able to access the information so much easier, no matter if you are accessing it via a browser (such as WAP) or via an application that is installed on the device. While traditional circuit-switched networks force users to go through a dial-up process, the connection will always be available for GPRS users. The difference is similar to using a PC with the dial-up connection of a modem (which is common in homes) and a broadband connection (which is dominant at workplaces and schools). The biggest difference (yes, even bigger than the difference in speed) is that the connection with the Internet over broadband is seamless, and you do not need a connection setup procedure. The network is always accessible (well, as long as the network is up and running, at least), and

Internet applications are as easy to use as those that run solely on the PC itself. We also illustrate this concept in Figure 3.3, where a GPRS packet-switched user and a GSM circuit-switched user are initiating a connection at the same time. While the GPRS user can start the session instantly and send and receive data, the CS data user has to wait for some time while the connection initiates. The use of packet data also opens new charging schemes where subscribers are charged more based on usage rather than on the duration of the connection. As you can see in Figure 3.3, there is no clear start and end to the GPRS session, because these sessions usually go on as long as the mobile is turned on.

In order to clarify this new feature, this example shows how the same application, getting a last-minute trip, changes with GPRS.

Home PC. You go home to your PC and check a Web site that has offers on cheap tickets. You connect, wait, and then check what is available and what the prices are. If you are in luck, the site might even have e-mail alerts that will tell you when a good deal occurs (most of us nonbroadband users will have to dial in every hour to check whether we have received that e-mail).

WAP-enabled GSM CS phone. With a circuit-switched WAP phone, you can access a similar site and check whether something interesting appears. You dial in and check to determine whether there are any offers (no matter where you are). You then repeat that procedure until you find something interesting (in other words, you hunt for the information and polls on a regular basis to update yourself on changes). Commonly, you would disconnect once you have checked and reconnect the next time. You could also make a regular phone call, but the process is the same.

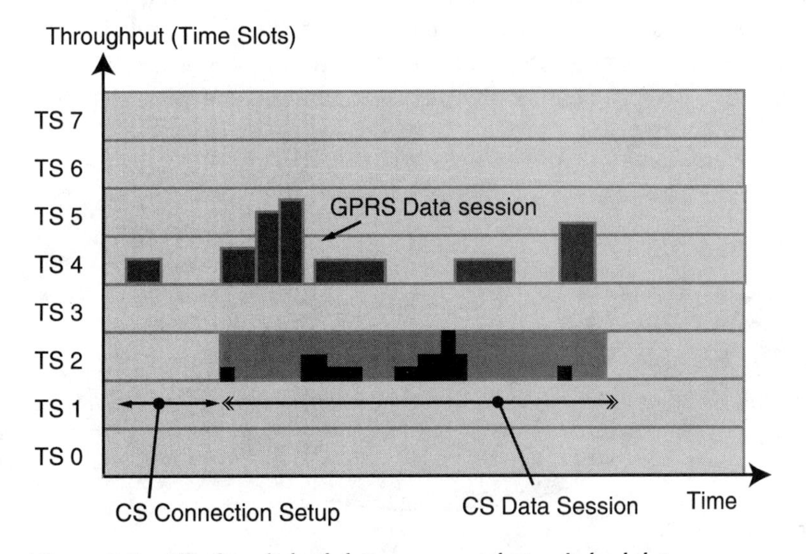

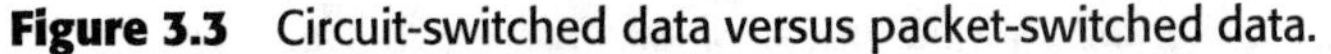
Figure 3.3 Circuit-switched data versus packet-switched data.

Cellular phone with GPRS and WAP. The GPRS phone is always online, so you can access the site with the tickets anywhere and anytime. You click on the kinds of preferences that you want (Hawaii, maximum price of $300, sometime in May, and so on). The application that handles the last-minute tickets then keeps those requests in mind and notifies you whenever something arises. Alternatively, you can have an application on your handset (provided that it uses an open platform) and have that application check for updates on a regular basis. You can then walk around relaxed, not bound to a single location or forced to check something on a regular basis.

As we illustrated in the previous example, GPRS moves into new usage patterns where the always online functionality makes it possible for our handheld devices to perform tasks for us. If applications developers can leverage this advantage, our handheld devices will be essential companions that make life easier for all of us.

An Upgrade to Existing Networks

GPRS is not a completely new system; rather, it is an upgrade that empowers existing GSM networks. In other words, you will still have the same functionality for voice calls, and it is even possible to have simultaneous voice and data on some handsets. This smooth migration also means that you will enjoy the same coverage for GPRS as for present cellular networks, as opposed to building a completely new network from scratch. This situation is possible because GPRS is introduced as a simple software upgrade for the majority of the operator's equipment: the base stations. In other words, most operators can upgrade to GPRS without having technicians traveling to each cell site. Instead, a centralized software update is possible. A large cost for today's mobile operators is also for the footprint of the base station and the antenna, and being able to reuse the same base stations saves lots of money and trouble. Figure 3.4 shows a simplified picture of how you can add GPRS on top of existing GSM systems.

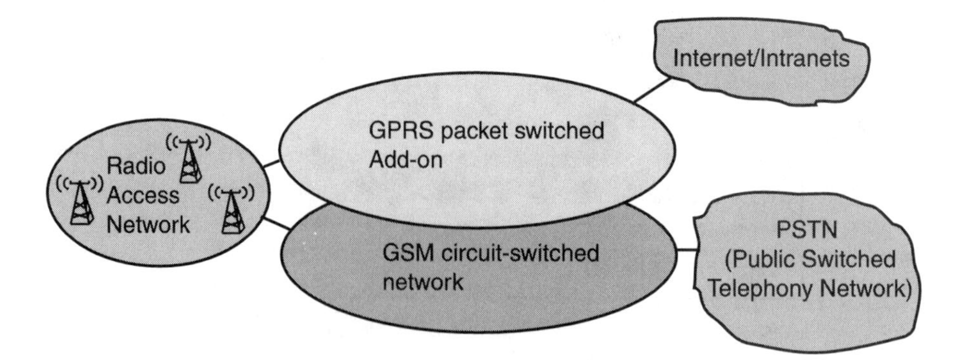

Figure 3.4 GPRS packet data added on top of existing networks.

Because the GSM network still provides voice and circuit-switched data functionalities, existing users will not experience a service degradation. That way, present phones will work in the future as well, but you will need new handsets in order to access the new packet data features.

An Integral Part of Future 3G Systems

Getting GPRS functionality into mobile networks will make people see some of the real benefits of the mobile Internet, but it will also create an urge for more: more speed, more capacity, and more features. Therefore, we predict that the introduction of 3G systems will occur in just a few years after GPRS. The 3G systems are, however, just another upgrade of the GSM/GPRS networks—regardless of whether EDGE or WCDMA is chosen. The GPRS core network handles the packet data and the always online functionality, and GPRS is then conveniently upgraded with the extra functionality. In Figure 3.4, we see that there are mostly changes in the radio access part and that EDGE/WCDMA radio access networks can coexist with GPRS/GSM networks. Some operators might choose to never upgrade the radio interface in rural areas but only offer GPRS/GSM service in those locations. The GPRS solution is highly effective in achieving high coverage, and you can then adjust 3G build-out according to customer demand.

The migration of GPRS to 3G is especially obvious when the second release of packet data core networks (the first one that supports 3G) from the major wireless vendors support GPRS and EDGE as well as WCDMA, and when the standard for the three is the same.

Extensive work in 3GPP and 3GPP2 harmonization groups has now also made it possible to mix core networks from one system with the radio interface of another system. In other words, a cdma2000 operator who has an ANSI-41 Mobile IP core network can run a WCDMA air interface, and a GPRS operator can attach a cdma2000 air interface. We describe the 3G systems in more detail in Chapter 4, "3G Wireless Systems."

GPRS Network Architecture

As we described previously, the added functionality of GPRS does not affect existing circuit-switched GSM services. We add the new packet data network as an overlay, where we reuse the existing infrastructure as much as possible. The starting point when looking at the architecture is the 2G GSM system, as seen in Figure 3.5.

In the figure, note that each MSC is connected to several BSCs, and each BSC in turn controls several base stations (BTSs). In order to simplify the picture,

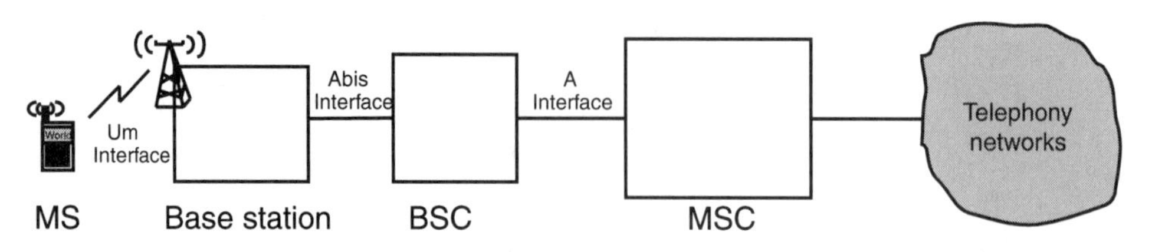

Figure 3.5 GSM system picture.

we have omitted some nodes. This figure includes the *Home Location Registry* (HLR), the *Authentication Center* (AuC), the *Equipment Identity Registry* (EIR), and the *Short Message Service Center* (SMS-C). These connect to the MSC via a *Signaling System 7* (SS7) network but will not play a central role in the discussions in this book.

One key force behind GPRS standardization is to make the transition as simple and as cost effective as possible. In other words, for instance, we should modify the base stations as little as possible. The base stations are, first of all, the lion's share of the equipment in which the operators have invested, and it is out of the question to replace them. Second, the base stations with their antennas are the elements of the network that create the coverage; thus, their deployment is spread out around the country in question. In order to facilitate maximum coverage, operators often place this equipment on rooftops and on hills, which makes it difficult and costly to perform on-site changes. A third and lesser-known reason is that the cell sites are often rented from the owner of the real estate (and, in some cases, from the tower's owner). The tower companies lease parts of the tower to different, often competing operators. Therefore, GPRS can be made as only a software upgrade (implementation specific; some have to do more) to existing base stations, which often can be done remotely from a central maintenance location. This software enables voice and data users to share the same air interface and to share base station resources, and it also makes it possible to develop new packet data-coding schemes. These coding schemes affect the resulting throughput of GPRS, and we describe them more in detail later in this chapter.

In GSM, the Abis interface is standardized to facilitate connectivity between multiple base stations and a BSC. This interface can remain unchanged when GPRS is introduced—again, to make the transition as smooth as possible. The data that goes over Abis consists of both GPRS packet data and GSM voice, because these components share the same air interface. In order to achieve efficient packet data handling, you need different core networks: the existing GSM core network for circuit-switched data and a new GPRS core network for packet data. We illustrate this concept in Figure 3.6.

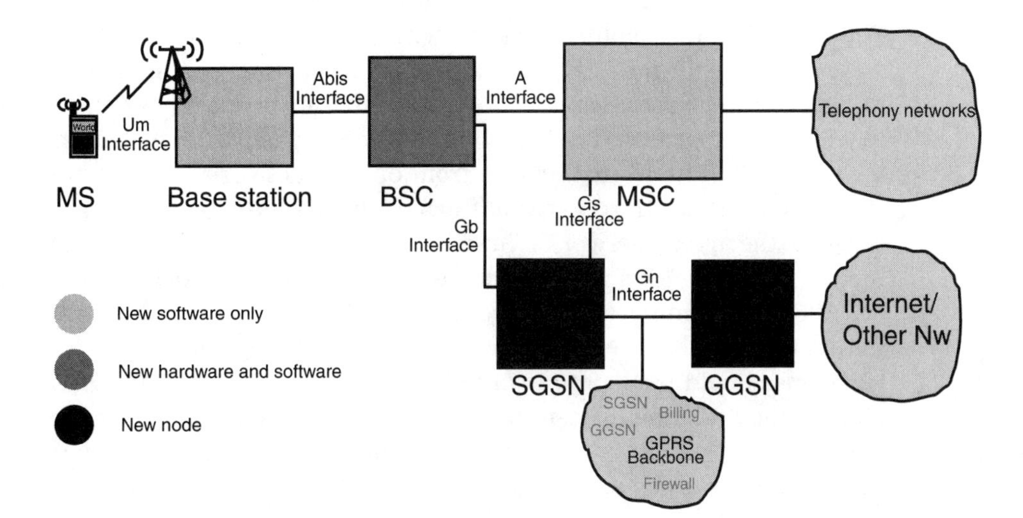

Figure 3.6 GPRS system architecture.

In other words, the BSC has to separate the different data flows and direct them to the right network. The additional functionality that it needs requires new hardware in the BSC: the *Packet Control Unit* (PCU). The PCU separates packet data and circuit-switched data when it is received from the MS and multiplexes the different data streams from circuit-switched and packet-switched core networks into common streams going down to the cells. The PCU is a separate entity and could potentially be located physically separate from the BSC. The BSC also gets its software upgraded for GPRS in order to enable it to handle the new logical packet data channels, the paging of GPRS handsets, and other packet data-specific functions of the air interface. Most of the new functionalities that we add to the GPRS air interface are thus implemented in the BSC. One BSC is connected to several base stations (varying from a just a few to hundreds of them per BSC), one MSC, and one *Serving GPRS Support Node* (SGSN).

The GPRS core network has two main nodes: the SGSN and the *Gateway GPRS Support Node* (GGSN), which together we call the GSN nodes. To connect these nodes to the radio network, a new open interface, Gb, is introduced. Gb is a high-speed Frame Relay link that is built running on an E1 or T1 connection. The connection between different GSN nodes and other components of the core network is called the GPRS backbone. The backbone is a regular IP network that has access routers, firewalls, gigabit routers, and so on. The backbone also usually connects to the operator billing system via a billing gateway (see the information later in this chapter). The backbone can also be used to connect to other GPRS operators.

The SGSN has the main responsibility for the mobility of packet data users. When connected to a GPRS network, the MS has a logical connection to its SGSN and can perform handover between different cells without any change in that logical connection. The SGSN keeps track of which BSC to use when sending packets to an MS that arrives from outside networks. Its functionality is similar to a regular IP router, but it has the added function of dealing with the issues of the mobile network. These issues include the authentication of users, the distribution of IP addresses, ciphering, and so on. The details of setting up a GPRS connection are described in the traffic cases subchapter as follows. Generally, a user will be within the same SGSN for long periods at a time, but if he or she should move into another SGSN service area, an inter-SGSN handover can be performed. Most of the time, the user will not notice this situation, although the packets that were currently buffered in the old SGSN might be discarded and re-sent by using higher layers.

Because the characteristics of a radio link are very different from those of a fixed link and bits over the air can be lost, some additional functionality is added. The RLC protocol operates between the MS and the base station and resends data that is lost over the air. The Logical Link Control (LLC) protocol, between MS and SGSN, can be configured to perform similar functionality. When an MS is connected to a site on the Internet, the majority of lost data will occur over the wireless link, and handling that with higher-layer protocols such as TCP would be very inefficient. You will find it much better to have a quick retransmission protocol that only covers the wireless part and hides the loss for TCP, which can then deal with the things for which it was originally designed.

The GGSN is similar to a combined gateway, firewall, and IP router. The GGSN handles interfaces to external IP networks, *Internet Service Providers* (ISPs), routers, *Remote Access Dial-In User Service* (RADIUS) servers, and other adjacent nodes. To the external networks, the GGSN appears as any gateway that can route packets to the users within its domain. The GGSN keeps track of the SGSN to which a specific MS is connected and forwards packets accordingly. The SGSN and GGSN can either be co-located in a compact GSN (CGSN) solution or placed far from each other and connected via the backbone. Because the backbone can be shared with other operators and with others (the operator decides on the architecture), a tunneling protocol called *GPRS Tunneling Protocol* (GTP) is used. In other words, packets that are traveling over the GPRS backbone have a stack with IP and TCP at two levels (see Figure 3.7). This procedure is not the most efficient, but it makes the solution secure and easy to implement. The majority of lower-layer GPRS protocols are not relevant to applications developers (operators typically configure them). Therefore, this book does not go into more detail about their functionality, and we explain the acronyms in the glossary.

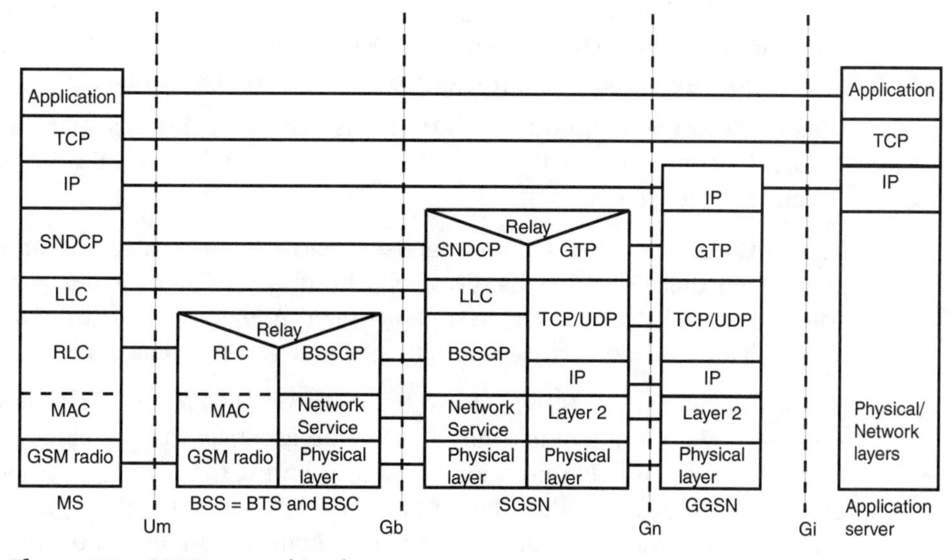

Figure 3.7 GPRS protocol stack.

Chapter 9, "Application Architectures," discusses how the GPRS network interfaces with the infrastructure that applications introduce and where to put applications servers.

GPRS Handsets

In order to take advantage of the new GPRS services, we need new handsets. In this subchapter, we emphasize the difference between the mobile terminal (the device on which the application runs) and the terminal equipment (the modem, here called the terminal). We can divide the terminals into three classes:

Class A terminals can handle packet data and voice at the same time. In other words, we need two transceivers because the handset has to send and/or receive data and voice at the same time. This situation makes class A terminals significantly more expensive to manufacture than class B and C terminals.

Class B terminals handle both packet data and voice, but not at the same time. In other words, you can use the same transceiver for both, keeping the cost of the terminals down. In practice, the GPRS session (like WAP browsing, file transfer, and so on) is suspended when a GSM voice call is started. How this information is presented to the user is up to the device manufacturer, but one way is to give the user the choice between receiving an incoming call and maintaining the data session. That way, a user who is transferring money between his or her accounts by using a WAP service does not have to stop that transaction just because someone calls.

Class C terminals can only handle either voice or data. Examples of class C terminals are GPRS PCM/CIA cards, embedded modules in vending machines, and so on.

Due to the high cost of class A handsets, most handset manufacturers have announced that their first handsets will be class B. There is currently work going on in 3GPP to standardize a lightweight A class in order to make handsets with simultaneous voice and data available at a reasonable cost.

The throughput of GPRS is said to be theoretically 170Kbps (probably on a sunny day with a nice tail wind). This value assumes that all eight available time slots are used, that no other users are sharing them, and that there is no protective coding. In reality, the terminal will often limit the throughput. There are 31 defined configurations of GPRS terminals that indicate how many time slots will be for downlink and uplink. Hence, a 4+1 terminal can receive data by using four time slots but can only send by using one. The capacity per time slot varies (9Kbps to 20Kbps) depending on the coding that you use, and we describe this coding later in this chapter. For GPRS phones, where limiting power consumption is a key success factor, this kind of asymmetric configuration is most common due to the heat generation and battery consumption that high data rates give. Sending data by using a higher speed requires more transmitting equipment. This transmitting equipment consumes electricity, which in turn is generated by using the battery. Also, the more energy that travels from the battery to the transmitter, the more energy is lost in the form of heat. Thus, it is difficult to build an 8+8 handset. The power consumption would be enormous, and the heat generation would require a massive cooling fan (which limits the practical usefulness of the device). Maybe it would be a hit in the northern parts of Sweden and Finland, where there is snow most of the time.

There are also physical and business limitations on the number of time slots that a multislot terminal is capable of using. If only one transceiver is used, which is likely to save cost and complexity, the mobile cannot send and receive at the same time. Initially, it is likely that terminals will be between 2+1 and 4+1 (perhaps 3+2). Most of these will be class B terminals. Mainstream terminals will unlikely be better than this value in the beginning. The reason is because there are only eight time slots per frame, which gives the terminal eight chances per frame to send four time slots and to receive one if a 4+1 terminal wants to achieve maximum speed.

Thus, a 1+4 terminal must be capable of using five time slots in that one frame for user data (1 + 4 = 5). That gives three slots left in the frame for the terminal

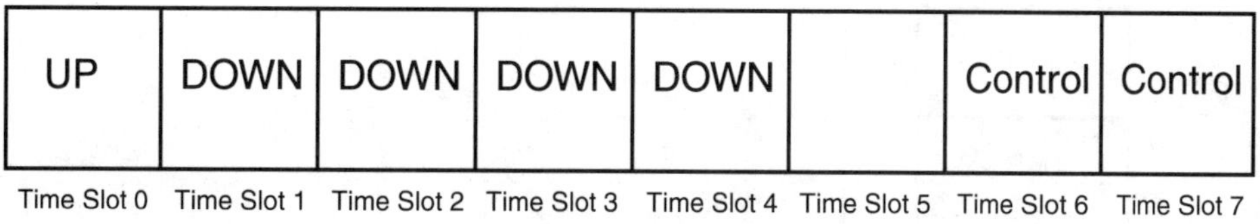

UP	DOWN	DOWN	DOWN	DOWN		Control	Control
Time Slot 0	Time Slot 1	Time Slot 2	Time Slot 3	Time Slot 4	Time Slot 5	Time Slot 6	Time Slot 7

Figure 3.8 Time slot usage for a one-transceiver terminal.

to make control signaling with the base station, to maintain a connection, and to monitor other cells to see whether a handover is necessary. We illustrate this procedure in Figure 3.8.

Now, imagine that we could make a 6+2 mobile and that we wanted to use it in order to achieve maximum speed. The eight time slots, where one time slot is mostly used for control data, would rarely be enough. Consequently, the terminal would have to have two transceivers in order to send and receive at the same time. This functionality adds a significant cost to the handset. Now, say that you want the same handset to be class A and to handle voice at the same time. This feature would require a third transceiver to be added, or the situation would be back at square one (where one transceiver is used for voice).

The R-Reference Point Interface

When the mobile terminal and the terminal equipment are physically separated, the R-reference point interface is used (as highlighted in Figure 3.9).

An example of such a configuration could be a Bluetooth-enabled laptop that communicates with network servers via a GPRS and Bluetooth-enabled phone. The phone then serves as a modem and acts like an Internet bridge (a modem) for the laptop. In order to establish such a connection, you first connect the devices on the physical layer. When a cable is used, this procedure is pretty obvious (devices are just hooked up to each other, and the serial ports are enabled). With infrared connections, the infrared hardware has to be enabled and the devices have to be held close enough and pointed toward each other. If Bluetooth is used, the devices are usually paired (we describe this concept in more detail in Chapter 5, "Bluetooth-Cutting the Cord!"). In any case, a physical connection is established between the TE and the MT. The next step is to set up a *Point-to-Point Protocol* (PPP) connection between the communicating devices. PPP is implemented in most operating systems, and the actual setup depends on the platform. PPP creates a communications interface toward the lower layers and enables the IP layer to seamlessly communicate with the network. In addition, it performs some optimizations such as avoiding sending

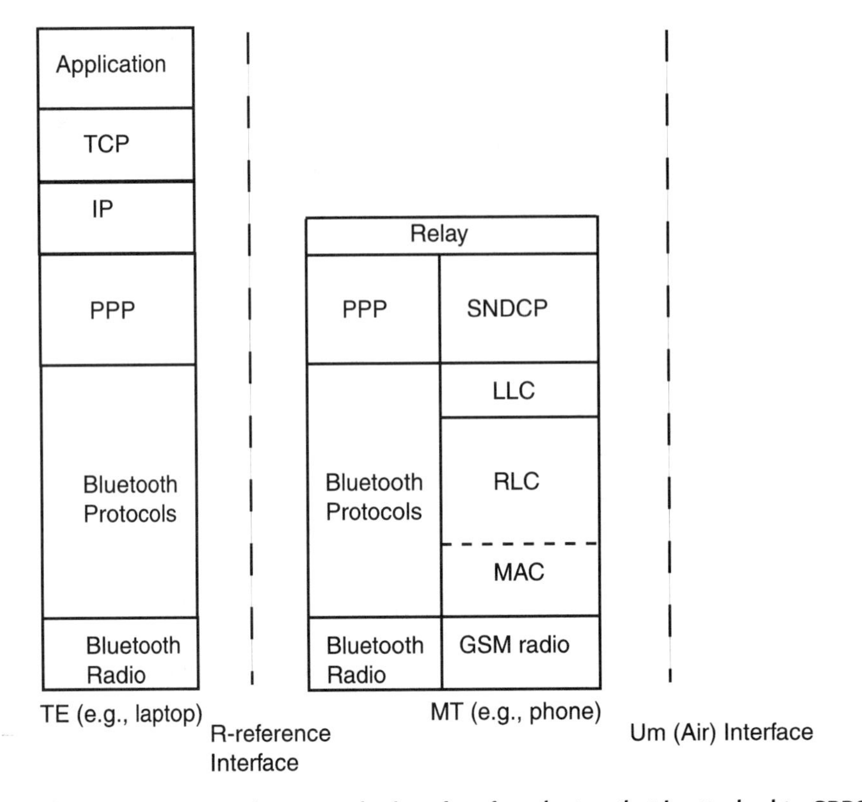

Figure 3.9 An R-Reference point interface for a laptop that is attached to GPRS via a Bluetooth-enabled phone.

redundant header information, and so on. Most dial-up connections today use PPP to connect to the access server, and most operating systems offer convenient abstractions for the user. In Windows 98, for example, there is a dial-up networking wizard that takes the user through a step-by-step process of setting up a PPP connection to the ISP.

Accessing Lower Layers with AT Commands

With the added functionality of GPRS phones, the interface from the MT to the TE becomes interesting. In order to make it possible for applications developers to access the GPRS terminal hardware and the network properties directly, the GPRS standard defines a number of *Attention* (AT) commands. The AT commands are developed with the ITU V.25ter recommendation (ITU-T Recommendation V.25ter: "Serial asynchronous automatic dialing and control") in mind that recommends a set of commands that communications equipment should offer to higher layers. Therefore, many of the GPRS AT commands are also seen in GSM and UMTS as well as fixed modems, and their technique can be applied generically. The ETSI standard GSM 07.07, Release 1997, describes the AT command set for GSM terminals, including HSCSD and GPRS. In the specifications, the MT is divided into a *Terminal Adaptor* (TA) and *Mobile Equipment* (ME). The idea is that the TA is the one that receives and interprets the AT commands, but in this book, we make no distinction between the two. Consequently, we keep the two parts (the MT, or the modem part, and the TE, or the applications platform part) in order to describe the handset, as illustrated in Figure 3.10. Note that it would also be possible to have the TA in the TE, but this situation will not be the case here. In Chapter 10, "Mobile Internet Devices," we give a more elaborate description about generic devices and how the MT and TE can be either separated physically or integrated.

In order to talk directly to the modem part of the MT, different methods are used ranging from system calls in the programming language to a simple Telnet session. For the sake of simplicity, no specific environment is used here; rather, we use a generic description of the AT commands themselves. Opening a Telnet session to the MT and typing the commands enables you to test these commands. Note, however, that many of the commands require the use of eight-bit (byte) mode.

The AT commands are entered either as basic commands, as defined in V.25ter, or as extended commands. GSM/GPRS commands use the latter format, and the syntax looks like the following:

```
ATCMD1 CMD2=98; +CMD3 <CR>
```

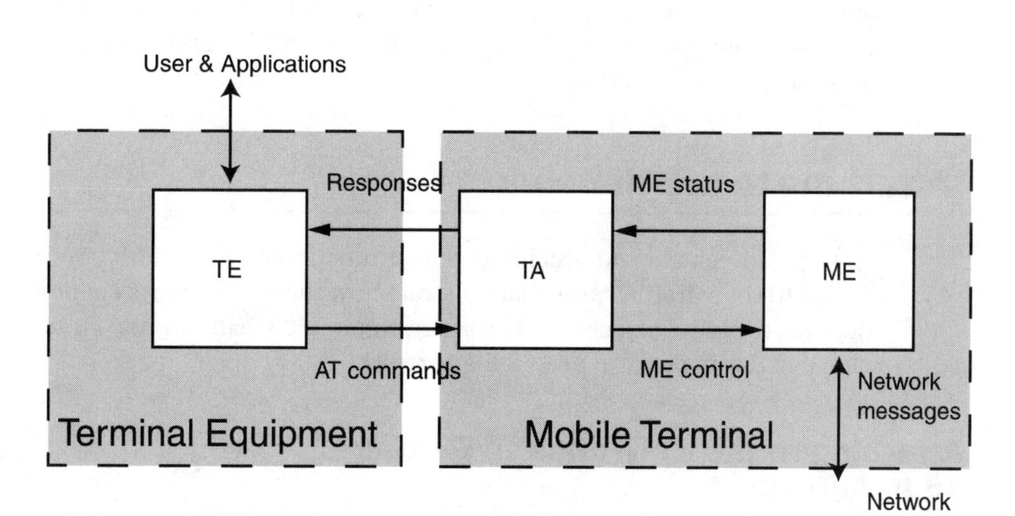

Figure 3.10 How AT commands access the MT.

The *AT* prefix initiates every command, and then multiple commands can be added with only a + prefix. In this example, *CMD1* is executed, followed by *CMD2* with the subparameter 12 and then *CMD3*. The *Carriage Return* (CR) symbol (<CR>) ends the command string. Now, let's take a look at a concrete example where we can check how much battery the terminal has left:

```
ATCBC <CR>
```

The battery charge command, *+CBC*, returns two values: battery connection status and battery connection length. The connection status tells whether the battery operation is used and attached, and the length indicates how much battery power the terminal has left (0 percent to 100 percent).

GPRS-specific commands also include the following:

- V42bis data compression on/off
- Header compression on/off
- Request QoS profile
- GPRS attach/detach
- Request class of operation (A, B, or C)
- Get GPRS network registration status

For detailed information about the syntax of individual commands, we recommend that you examine the standard document GSM 07.07.

For information about how to access these commands on individual handsets, please refer to the *Software Development Kits* (SDKs) that the device manufacturers provide. SDKs for Ericsson, Nokia, and Motorola phones are available on their respective developer sites. Links can be found on the Web site for this book, see Introduction. In addition, most operating systems have a way of accessing AT-commands.

Attaching to the Network

In order to cement the architecture and functionality of GPRS, we will now look into some traffic cases that illustrate how the system works in practice. In the first scenario, a user who has an Ericsson R520 feature phone turns on the phone and gets ready to access the network.

Attaching to the Network and Getting an IP Address

In this case, the MT and TE are both inside the phone. The R520 is a combined GSM and GPRS phone (class B). In other words, it needs to tell the network that it can make and receive both GSM and GPRS connections. This procedure is called Attach and is similar for GSM and GPRS. We use GSM here to indicate the circuit-switched (voice or data) sessions. Similarly, GPRS is synonymous with packet data sessions. Here, we do not describe in detail the International Mobile Subscriber Identity (IMSI) Attach that registers the GSM part; instead, we refer interested readers to the GSM standard. This process includes authentication, ciphering, and so on and puts the user in an idle state (ready to make or receive calls).

Performing a GPRS Attach creates the logical link between the SGSN and the MS. This task is done in the following way:

1. The MS sends an Attach request message to the SGSN.
2. The SGSN checks to determine whether it knows the MS and tries to find its unique IMSI identification number. If the MS is not known, it asks the old SGSN for IMSI and authentication triplets.
3. If the old SGSN does not know the MS, it sends an error message. The new SGSN then asks the MS for its IMSI. One would think that it would be more efficient to ask the MS right away, but sending the unique IMSI number over the air is generally avoided for security reasons.
4. The SGSN performs an authentication of the MS.
5. If the MS is found to be in a new service area, the Home Location Registry (HLR) is updated.
6. If the MS currently is in a new location area, the Mobile Switching Center (MSC)/Visitors Location Registry (VLR) is updated.
7. The SGSN tells the MS about its assigned *Temporary Location Link Identifier* (TLLI). TLLI is used throughout the GPRS session as an identifier for the MS-SGSN logical link.

It is possible to perform a combined GPRS/IMSI Attach and make the phone visible to both voice and packet data at the same time. The opposite of a GPRS Attach is a GPRS Detach, which removes the GPRS terminal from the network. We will not describe this concept further here, but this procedure is usually done once the phone is turned off (just like the Attach is commonly done when the phone is turned on).

Now that the MS-SGSN link has been established, the mobile needs to get an IP address and other connection parameters. This task is done through *Packet Data Protocol* (PDP) context activation. The PDP context can be viewed as a software record that holds parameters that are relevant to a certain connection. This information includes the protocols that are used (IP or X25), the IP address (if IP is used), the QoS profile, and information about whether to use compression. The desired PDP context parameters can be set by the application using AT commands, as we described in the previous section. The PDP context activation makes the GPRS mobile visible to the concerned GGSN, which makes external connections possible. The following steps illustrate the PDP context activation procedure (note that a GPRS Attach has been performed previously):

1. The MS sends a PDP context request to the SGNS.
2. Security functions can be executed between the MS and the SGSN, which validates the request.
3. The SGSN:
   - Checks the subscription
   - Checks the QoS, which affects the pricing of the service
   - Sends information to the GGSN about how to reach the MS
   - Configures a logical link to the GGSN by setting up a tunnel
4. The GGSN contacts a RADIUS within the operator network and gets an IP address for the MS.
5. The IP-address is sent back to the MS.

A GPRS Attach followed by a PDP context activation is shown in Figure 3.11. In the figure, the TE could be a laptop, and the AT command is initiated when the user clicks a GPRS service icon.

Now, the GPRS user is ready to send and receive packets. You should note that this method is only one of the ways that an MS can get its IP address. The following alternatives are available:

Dynamic IP address. The GGSN assigns a dynamic IP from its own pool of IP addresses.

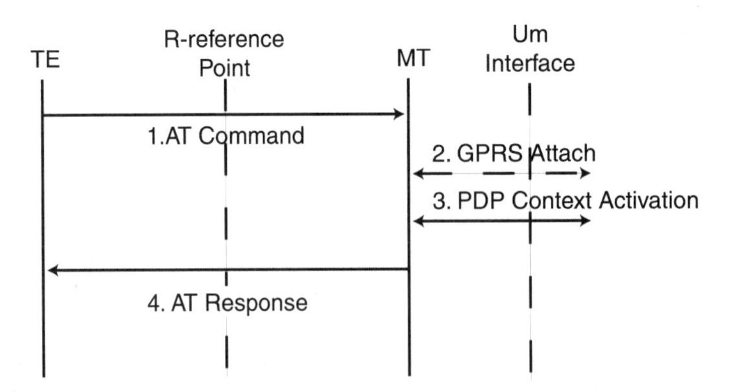

Figure 3.11 A laptop (TE) connects to the GPRS network via AT commands.

Dynamic IP address. The GGSN asks an ISP for an IP address, which usually is done via a RADIUS server. The ISP can be the mobile operator, as in the previous example, or it can be any external party.

Static IP address. The SGSN gets the IP address from the HLR (just like any other subscriber information).

Rarely will you use static IP addresses, because the use of IP version 4 already puts a strain on the supply of public IP addresses. The successor of IPv6 is being adopted gradually within the Internet community, but it will take time until all nodes are upgraded in order to take advantage of this technology. IPv6 is not supported in the initial release of GPRS, but we expect it in later releases. GPRS and the mobile Internet are likely to be the main drivers for widespread IPv6 adoption. There will just be too many mobile Internet users for IPv4 to handle in the long run.

Mobility Management

As a GPRS user moves around, the changes that are made to the connection must be seamless for the user. The first part of this goal involves soft handovers, a make-before-break feature that makes the handset connect with the cell that it is entering before dropping the cell that it is leaving. The mobility management principles are the same for GSM and GPRS as long as a user is within the same SGSN service area. The BSC ensures that the data is transported to and from the right cell, and the packet handling is not affected as the radio part tunes in to the new antennas of the cells that are entered.

Saving power is crucial for handheld devices—especially those that are connected to the network at all times. In order to save power, a GPRS terminal can transition between different modes of activity. When it attaches to the network, it changes its state from Idle to Ready. In this state, it can send and receive packets instantly. If no packets are sent or received for a period of time, a timer is triggered and the mobile enters Standby state. The device then stays in that state until packets are sent or received or until another timer is triggered and sends it back to Idle state. Some operators might choose to set this parameter (timer2 in Figure 3.12) to infinity in order to avoid users being forced to detach because of staying inactive too long. We show the different states and transitions in Figure 3.12.

The point of these different states is that they enable the MS to keep its battery consumption down when it is not actively communicating. These states minimize the communication with the system whenever possible. When a mobile is in Ready state, the SGSN needs to keep track of it on a cell level in order to know where the packets should be sent. In Standby state, on the other hand, the mobile is not currently sending or receiving data and has been inactive for some time. In other words, the SGSN does not have to know exactly where it is; rather, it just needs a rough idea of where it can find the mobile. This area is

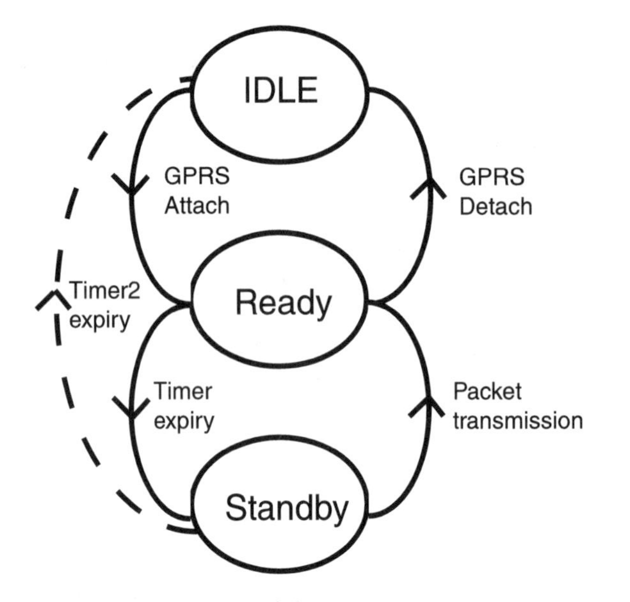

Figure 3.12 GPRS mobility states.

called a routing area, and within this area, a Standby MS can move around without needing to send any updates to the network (thus minimizing power usage). As the MS moves into a new routing area, it sends an update to the SGSN, letting it know where it can be found. If the MS initiates a packet transmission, it is moved into Ready state and again lets the SGSN know when it is changing cells. When the SGNS wants to reach a mobile that is in Standby state, it has to page the mobile device. GPRS paging is similar to the paging that 2G systems use when someone wants to call you on your mobile phone. The base stations in your area shout that they are looking for you, and your mobile replies. Because a Standby user is not known on the cell level (but is known on the routing area level), the system needs to page the user in the entire routing area (which might create an additional delay of a few seconds). This situation is worth considering when you are designing applications that seldom communicate with the user, because paging will most of the time be necessary in order to get the packets delivered. The entire routing area concept was created in order to make the area of paging as small as possible but to still enable the mobiles to save power. The circuit-switched equivalent is called a location area, and those areas are generally significantly bigger.

Communicating with a GPRS User

Now that we have seen how a GPRS system works, let's examine some applications that work in GPRS. First of all, GPRS applications are mostly IP applications. That is, their behaviors are more similar to an application for a *Local Area Network* (LAN) than some of the wireless networks that exist. You are not calling a GPRS user or sending information to phone numbers; rather, you are sending IP packets to a host. In addition, the GPRS networks have gateway and proxy functionalities in the GGSN, which turns them into something like a corporate intranet with protected, private IP addresses. Once a GPRS user has connected and has received his or her IP address, he or she is ready to start sending and receiving data. The IP address is hidden most of the time from the outside world (private), and therefore it is hard to start communicating with a GPRS user without the user taking the initiative. Most GPRS applications are client/server based, where the client sends requests and receives responses (much like the usage patterns on the fixed Internet). So, how can we initiate a transmission to a user?

First of all, if the user is not attached to the GPRS network, you cannot send packets to him or her. A function called network-initiated PDP context activation, where the network wakes up a mobile and attaches it to the network, is not supported in the initial release of the standard (release 1997). Pushing data from the network to a user can be done in two ways:

- By using WAP push, any WAP 1.2.1-compliant device can be configured to accept pushed information from the network. The user can then choose the sites from which he or she will accept pushed information.
- For those devices that use an open platform, such as EPOC, a client can be created that listens to a socket and that has a link to the server in question. Instant messaging products such as America Online's Instant Messenger, ICQ, and IPulse are good examples.

Generally, if you can make an application work on a LAN where the client is behind a proxy or a firewall, you can easily make it work in GPRS and on other mobile Internet networks. The challenge, then, is to make it robust and efficient enough to work with the different characteristics of the wireless network. We describe how to perform this task in Chapter 8, "Adapting for Wireless Challenges."

Speed, Capacity, and Other End-User Aspects

Two common questions are, "What is the maximum speed of GPRS?" and "What is realistic to see in practice?" The main factors that decide the speed are as follows:

- The number of other packet data and voice users in the cell
- The number of time slots that the handset supports
- The coding scheme used

Coding of the bit stream over the air is done in order to give different degrees of robustness. The coding makes it possible to recover data packets even if a few bits are lost on the way. For GPRS, four coding schemes are defined (CS-1, 2, 3, and 4), and as shown in Table 3.1, they significantly affect the resulting throughput.

The table shows speeds at the LLC level while the physical speeds will be 9,050 bps, 13,400 bps, 15,600 bps, and 21,400 bps, respectively.

The aim is to use a more robust coding when the signal quality is low (CS-1) and to use less robust coding in better radio conditions (CS-4). The network chooses the coding scheme, and some systems initially only support CS-1 and CS-2. As we saw previously, the maximum speed for an eight-time-slot mobile that uses CS4 would be 160Kbps. A more realistic maximum calculation uses CS-2 and a four-time-slot mobile, which results in 48Kbps. This can be improved further by using compression technologies such as v42bis.

When configuring the GPRS network, the operator can choose how much capacity will be available to GPRS packet data and how much the circuit-switched data and voice users should receive. A common way of configuring the cells is to make time slots available On Demand. In other words, these users can use them for packet data as long as no circuit-switched traffic occupies them. Once a circuit-switched user wants to connect, the packet data users are pre-empted from that time slot (note that each HSCSD user can only pre-empt one time slot). Another configuration option is to make some Dedicated GPRS time slots, which reserves them for packet data only. Considering that GPRS is an upgrade to existing networks where there is usually already a lot of circuit-switched traffic, this issue is sensitive. Dedicating time slots to GPRS in reality gives less capacity for the existing customer base and increases the potential revenue loss. Therefore, many operators are reluctant to allocate dedicated GPRS time slots and prefer the On Demand alternative. As traffic increases, they can then build additional capacity by installing more base stations. In simulations and GPRS trials during 2000, it has been shown that the GPRS throughput will be low if no time slots are dedicated. In those cases, the handset capacity will rarely be the limitation—and throughputs will remain lower than 10Kbps even with a moderate GPRS load. Operators will have the

Table 3.1 GPRS Coding Schemes

CODING	MAXIMUM DATA RATE (KBPS)	TARGET C/I (DB)
CS-1	8	-6
CS-2	12	-9
CS-3	14	-12
CS-4	20	-17

challenge of meeting the demand for capacity, because many of them get so many new subscribers just for 2G voice that it is tough to build out the capacity fast enough.

Because the number of users affects the performance significantly, it is (of course) interesting to determine how many users there will be per cell. Estimating this value without knowing how rapid the takeup of GPRS will be is difficult, but there is likely to be as many as 30 to 40 users sharing the same eight time slots at times. This number sounds worse than it really is, because a typical packet data session includes lots of inactivity. The most significant aspect of GPRS throughput is that you can never take it for granted. One second you might have 40Kbps, and the next second you might be pushed down to 5Kbps. Likely, the downlink bit rate for a 4+1 phone will be in the 5Kbps to 40Kbps range most of the time.

QoS is another area where there is much confusion. This situation is understandable, however, because it is a bit complicated and not that well documented. Keeping the hardware and software changes to a minimum while introducing advanced control mechanisms in order to ensure quality for individual users is not an easy task. The QoS architecture also needed to be flexible enough to accommodate the high demands that would be placed upon UMTS. The result is that there are a number of parameters that can be set, but GPRS QoS will be best effort most of the time (mostly due to limitations of the PCU, which we could not overcome without substantial changes to the architecture). Simulations have shown that the SGSN is perfectly capable of doing the differentiation, but the SGSN's efforts are diluted by the PCU's inability to do so. Therefore, do not count on QoS features being available for your applications in the first release of GPRS.

Charging for Packet Data

As packet data is introduced into mobile systems, the question of how to bill for the services arises. Always online and paying by the minute does not sound all that appealing, does it? The question of how billing in GPRS will work is not primarily decided by what is possible but relates more to what the operator chooses. Here, we describe the possibilities and give some examples at the end.

The SGSN and GGSN register all possible aspects of a GPRS user's behavior and generate billing information accordingly. This information is gathered in so-called *Charging Data Records* (CDR) and is delivered to a billing gateway. The charging can be based on the following parameters:

Volume. The amount of bytes transferred.

Duration. The duration of a PDP context session.

Time. Date, time of day, and day of the week (enabling lower tariffs at off-peak hours).

Final destination. The destination address (a subscriber could be charged for access to the specific network, such as through a proxy server).

Location. The location of the subscriber. Perhaps hotspots with different tariffs could be a way to charge the end user. Another possibility is to have differentiated tariffs and to have lower costs in home areas compared to office areas and enabling home zone and office zone concepts.

Quality of Service. Pay more for higher network priority.

SMS. The SGSN will produce specific CDRs for SMS.

Served IMSI/subscriber. Different subscriber classes (different tariffs for frequent users, businesses, or private users).

Reverse charging. The receiving subscriber is not charged for the received data; instead, the sending party is charged (or perhaps a third party).

Free of charge. Specified data to be free of charge.

Flat rate. A fixed monthly fee (we believe that this method will be the preferred way of charging because it will attract the mass market).

Bearer service. Charging based on different bearer services (for an operator who has several networks, such as GSM900 and GSM1800, and who wants to promote usage of one of the networks). Or, perhaps the bearer service would be good for areas where it would be cheaper for the operator to offer services from a wireless LAN rather than from the GSM network.

There is no functionality within the GPRS core network to charge based on what application is used. This feature can be added to the mobile network through a payment server on the service network (see Chapter 9, "Application Architectures").

Now that we have seen what you can charge for, it is up to each individual operator to decide on what solution he or she wants for the subscribers. One possible scheme is to sell subscriptions with a flat monthly rate, as long as the data usage does not exceed a certain limit. This scenario is fairly simple to implement, and users should understand it pretty easily—although the notion of paying per packet or kilobyte will sound strange to most people. Another scenario is that an *Application Service Provider* (ASP) will buy capacity from the operator and will offer a service toward its customers that includes free data usage. This situation is only possible with those applications that are low on bandwidth usage (such as WAP applications). Likely, we will see a tendency to charge more and more for the value-added services than the actual usage as we move into the future. It is doubtful, however, that bandwidth will be truly free, because the spectrum that is used in the air is expensive. Different models might arise, though, that will relieve the end user from the cost in exchange for advertisements or promotions.

The Future of GPRS

We have now seen that GPRS is a crucial step in the mobile evolution, and it opens endless possibilities for application developers and users. The next step after GPRS can be either EDGE or UMTS (or both). As we mentioned in the previous chapter, GPRS will be introduced into TDMA systems as a part of the EDGE upgrade and not as a separate system. This step is important, because it makes it possible to build handsets that work on both TDMA and GSM networks.

One of the most important additions to GPRS in the second release of the core network (3GPP release 1999) is the extensive QoS functionality. This standard is the same for EDGE and UMTS and specifies different QoS profiles and their parameters. Another important part of this new standard (which, of course, is backward-compliant) is the possibility to use several services for one MS and have different qualities of service for them. We describe these features in more detail in the next chapter, where we give an overview of the different 3G systems.

Summary

GPRS is a packet data overlay system that upgrades existing networks. GPRS introduces three key features: always online, a convenient upgrade with instant coverage, and a road map to 3G. GPRS is a new core network that enables wireless packet data. The application developer can access the advanced features of GPRS by using AT commands, and GPRS turns the handset into an IP-based device on which just about any Internet application can run.

CHAPTER 4

3G Wireless Systems

As you have now seen, the 2.5G systems such as *General Packet Radio Services* (GPRS) give us many new possibilities when we are developing applications. Also, it is obvious that new needs now exist and that people in general want fancier handsets and more network capabilities. *Quality of Service* (QoS) for applications was clearly not prioritized in 2.5G, and the bit rates are not all that high in reality. Also, items that are not intimately connected to the standard itself, such as bigger displays and more processing power in the handsets, are all items that regular users crave. Other important features of the systems that, in most cases, only application developers and operators see missing are open *Application Programming Interfaces* (APIs) and architectures. We address most of these issues as we take the next step toward mobile evolution: third-generation wireless systems (known as 3G).

What Is 3G?

Again, as new systems emerge, it is crucial to have a thought-through migration strategy. In other words, 3G systems (most of the time) include 3G, 2.5G, and 2G functionalities. All of the alternative 3G standards will have seamless handover capabilities for 2.5G and 2G from day one—in networks as well as for handsets. That said, an operator initially can cover urban areas with these new high-capacity networks and cover the rest with 2G and 2.5G networks. For example, in countries where *Wideband Code Division Multiple Access* (WCDMA) is deployed, users who move into rural areas will initially be handed to a GPRS network if they are using packet data and to GSM if they are talking.

We often hear 3G being discussed synonymously with higher speeds, but what is a 3G system by definition? The *International Telecommunications Union* (ITU) has made a recommendation (ITU-R M687-2) on what the 3G systems, or *International Mobile Telecommunications 2000* (IMT-2000), should bring. This recommendation includes the following items:

- A QoS that is comparable to fixed voice networks
- A phased development, with the first phase supporting bit rates of up to 2Mbps
- The capability to build terminals that have many different form factors ranging in size from what 2G phones offer up to what you can carry in vehicles
- A flexible architecture where you can easily add additional applications

The recommendations, of course, included many items and were fairly general. The involved companies agreed that things such as flexible multimedia management, Internet access, flexible bearer services, and cost-effective packet access for best-effort services were of high importance. Because the Internet has become a global force and a daily tool for people (both professionally and privately), it is important to define a wide-area wireless system that is capable of utilizing all of those services. The challenge was to migrate toward an architecture where all of the benefits of the Internet remained while still preserving the high QoS of 2G systems (with low down times and guaranteed bit rates in 2G via circuit switching). The vision of a mobile Internet where not only Internet services but also a whole new range of tailored services would emerge started to form. This vision included the capability to access the services any time, anywhere, and on any device.

The ITU also arranged a conference in 1992 to determine which frequency bands it should recommend for 3G. The meeting that gave name to the following recommendations was the *World Administrative Radio Conference 1992* (WARC-92). The ITU identified the frequencies around 2GHz as suitable for use both for satellite and terrestrial mobile systems. The original target was to have a single 3G or IMT 2000 standard, but this goal was, as we will soon discover, very difficult.

During the late 1990s, there was an intriguing race between a number of camps in order to convince the world that their idea of 3G was the best. The different contestants in this race all had reasons for liking some proposals better than others (such as patents, in-house competence, similarity, compatibility with legacy systems, and so on). At the same time, everyone had the feeling that things needed to work better than in the 2G systems, where the different and incompatible standards made international roaming difficult and expensive.

In 1997, this standardization was driven separately in the United States, Japan, and Europe, although the participating companies often were present in all of the standardization bodies. In the first half of 1998, Europe made several decisions in the direction of WCDMA while the United States supported EDGE and cdma2000. Japan was also working toward standardizing WCDMA, but there were some key differences between its work and the European standard. In 1998, the ITU called for proposals for IMT -2000, and 10 proposals were submitted for the terrestrial part (the satellite part is not covered in this book). These proposals spurred several standards to work toward harmonization and the Japanese standardization body, ARIB/TTC, and the European counterpart, ETSI, T1P1 (United States), and TTA (Korea) to join forces in the strive toward a global standard. The result was one WCDMA standard, and the *Third-Generation Partnership Project* (3GPP) formed. U.S. standardization bodies then created 3GPP2, which standardizes the cdma2000 system. Also in 2000, GERAN (GSM EDGE Radio Air Interface) was added to 3GPP.

After additional harmonization work resulted in compromises between the different CDMA standards, they became closer to each other but still had three modes of the CDMA standard (as we will describe later in this chapter). In addition, EDGE is also part of the IMT 2000 family of 3G standards. The work toward making the different standards compatible is an ongoing process, and it will probably take some time. Not only does it involve technical issues, but it also involves the business aspects for operators who have customers who have legacy handsets to consider. The good part is that most applications will run on top of the *Internet Protocol* (IP) over any of these bearers, making it easy for developers to produce products that work anywhere.

Key Features of All 3G Systems

Before looking at the different 3G technologies in more detail, we will gather information about a number of key features that are common to all of the systems. We can then use this lowest common denominator when we design applications that will run across all networks. After all, most of the applications that we will see are plain IP applications that work independently of the bearer. Understanding the wireless systems, however, makes it possible to leverage its features and to overcome its challenges. In this chapter, we will not discuss the devices in-depth; rather, we will examine only those new aspects that the 3G systems add. In addition, advances in chip, battery, and display technology will make the devices even more advanced, but we will discuss these topics in Chapter 10, "Mobile Internet Devices."

Higher Bit Rates

Although the 2.5G systems introduce higher bit rates than 2G, it is not something that a user can count on. Users share bandwidth, and an application cannot take speed for granted. As more and more users get used to accessing the fixed Internet with broadband connections at home, the speed of mobile Internet access also becomes more important. More advanced devices that have bigger screens create an urge for fancy graphics and rich media, which requires higher speeds in order to be delivered. With 3G the speeds will become higher, and it will not be uncommon to get hundreds of Kbps during both uplink and downlink. You can achieve these speeds for channels that are either circuit switched or packet switched, though. For packet-switched channels, the resulting bit rate is highly dependent on the chosen QoS.

QoS

When a connection is set up between a user and the network, an agreement comes into place between the user and the operator that is dependent on the user's subscription. This agreement states what kind of delays the user should expect, what bit rates he or she should expect, and so on. For high-end subscriptions, a user might be guaranteed that the bit rate will never go lower than a certain number (unless he or she moves out of the coverage area, of course). In WCDMA and cdma2000, developers included the QoS aspects of the systems from the beginning so that the systems would support QoS end to end. As we saw in the previous chapter, GPRS has difficulties guaranteeing any QoS because of a lack of support in the Base Station Controller (BSC). This situation is not the case for EDGE , WCDMA, nor cdma2000.

Bit Rates Dependent on Distance

The third common feature of all of the 3G systems is that the maximum bit rate will be highly dependent on the distance from the base station. The reasons for this situation are a bit different in *Code Division Multiple Access* (CDMA) systems (cdma2000 and WCDMA) and in TDMA systems (EDGE), but for the user, it will appear similar: The further away you move from a base station, the harder it is to achieve high speeds. To some extent, the QoS management can remedy this situation, but there are physical restrictions that limit the speed at all times. The following descriptions of the individual systems go into more detail about these limitations. In addition, the 3G systems will often consist of different technologies within the same network. For instance, for WCDMA, coverage has to be built from scratch—leaving rural areas uncovered in early phases. In those areas, GPRS can be used as an alternate solution, and the WCDMA handset can perform a handover to GPRS when it runs out of WCDMA coverage. This process places new requirements on applications because they should be capable of functioning even as the bit rate goes down (maintaining some basic functionalities).

Similarly, in a cdmaOne network, cdma2000 coverage is likely to be built out initially in urban areas, where capacity and demand for higher data rates is prevalent. Like the GPRS/WCDMA scenario that we mentioned previously, subscribers who roam from cdma2000 coverage to cdmaOne coverage might experience a degradation in the QoS. This situation implies that developers, regardless of the underlying infrastructure, need the same kind of awareness.

Layered Open Architecture

As the services of the mobile Internet become more and more advanced and many ways of building the networks emerge, it is crucial to have a flexible architecture. In 3G, this is achieved by dividing the core network into a transport layer, control layer, and an applications/service layer, as shown in Figure 4.1.

Applications/service layer. This layer is where the applications are hosted and supporting services are offered. This layer is often called the service network, as we will describe later.

Control layer. Setting up calls, authenticating users, and making sure that all of the system intelligence works smoothly in the mobile network are tasks of the control layer. This functionality resides primarily in the traditional network nodes, the RNC/BSC, the MSC, the SGSN and the GGSN.

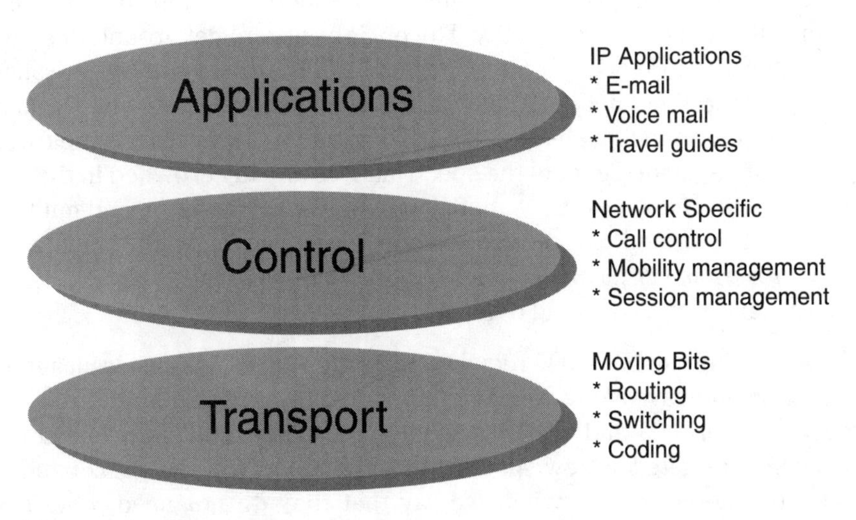

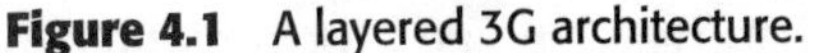

Figure 4.1 A layered 3G architecture.

Transport layer. The Transport layer transports the physical bits over the IP backbone and the wireless access network, regardless of whether the network is *Asynchronous Transfer Mode* (ATM), SONET or something else.

By using this architecture, an application in a service network (Application layer, described in more detail in Chapter 9, "Application Architecture") can be accessed from a user on the fixed Internet as well as from a 3G handset. The application server and the core transport network can be the same. As the data comes closer to the user, it is routed to the right access network (fixed or wireless). This feature is a significant advantage for application developers who can develop applications that work independently of the underlying bearer. Almost any IP-based application will run on the 3G networks. In addition, there will be APIs between the different layers, as shown in Figure 4.1. In other words, the functionality of underlying layers can be accessed through open and standardized interfaces.

New Spectrum?

When looking at the individual 3G technologies, you will find that the main difference is the use of the spectrum (or frequency). Many of today's 2G operators are struggling with capacity issues due to the phenomenal growth of cellular telephone users. When you are moving to 3G, getting more capacity is for some almost as important as obtaining new, fancy services. An operator always has a limited amount of frequency space (or spectrum), which limits the number of simultaneous subscribers to the network (see Figure 4.2). No matter what technology is used, going above this limit causes increasing interference between users and results in lower quality. The obvious way to get around this problem is to allocate more spectra for the operators, but there are some issues surrounding that. First, networks and handsets have to use the same frequency in order to work together, so the standard in question always limits the frequency choices of a system. Second, the spectrum was already crowded in the 1990s—not only with cellular systems, but also with television, radio, and military systems. In the United States, 2G systems already occupy the 1900 *Personal Cellular System* (PCS) band, which makes auctions of the 2000 MHz band difficult (because the two would overlap).

When defining the IMT-2000 requirements, the ITU took these issues into consideration and gave some choices of technologies that did not all require new spectra. Cdma2000 and EDGE can be deployed without new spectra, while WCDMA is targeted at new spectra (most commonly in the 2GHz band). Note that we generalize a bit when we say that they do not need more spectra, because the added capacity need often creates a need for replanning the networks (even for cdma2000 and EDGE).

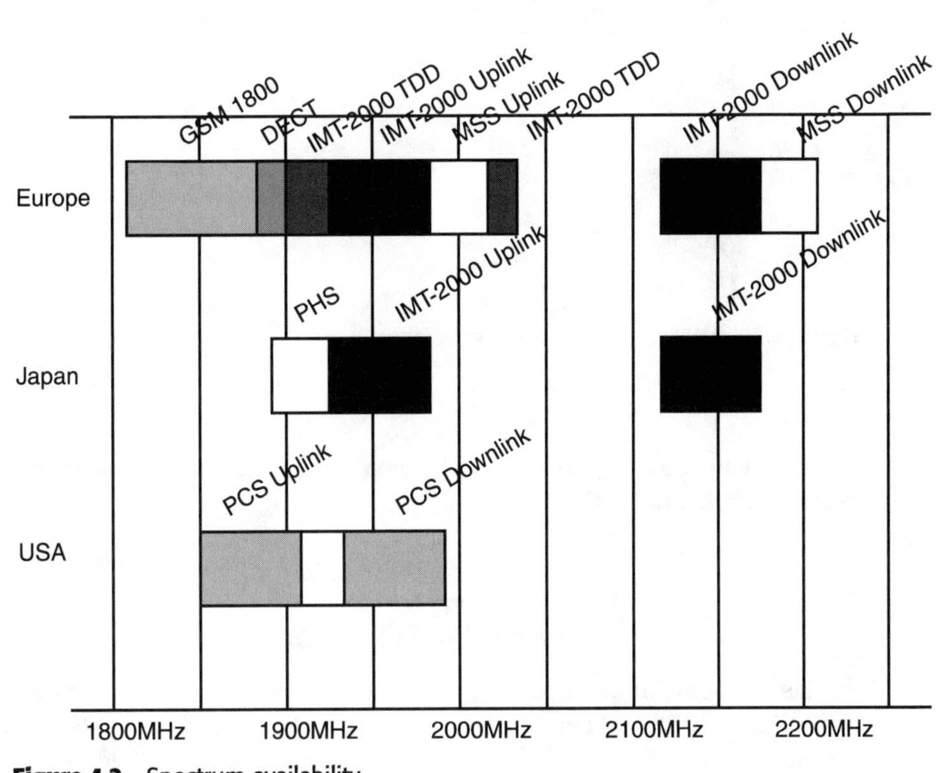

Figure 4.2 Spectrum availability.

System Description: EDGE

Enhanced Data Rates for Global Evolution (EDGE) or Enhanced Data Rates for GSM Evolution, which it was originally called, is a cost-efficient upgrade to existing GSM/GPRS and TDMA networks. EDGE operates in existing spectra and boosts the speed over the air interface. You can think of EDGE as a mechanism that squeezes in more capacity into each resource (time slot) over the air interface.

As we introduce GPRS into GSM networks, we can offer packet data services, and the usage becomes much easier. HSCSD makes it possible to use several time slots for each user, bringing higher data rates to circuit-switched services. Now that these two services are available, there is a need for even higher speeds for both of them. This feature is exactly what EDGE provides, making it possible to transfer more data in each time slot. While a GPRS upgrade mainly consists of new nodes in the core network, EDGE accelerates speeds over the air. When sending data over a wireless link, bits are commonly coded into symbols (in other words, some representation of information that can be sent over

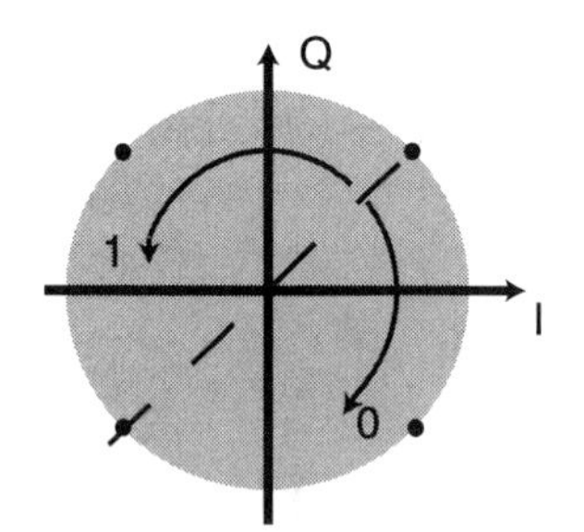

Figure 4.3 GMSK coding; one bit per symbol.

the air). One way of performing this coding is to let one bit correspond to a phase shift in a radio signal.

Bit	Phase shift
0	send: 315 receive: 225 < Phase < 45
1	send 135: receive 45 < Phase < 225

In this way, one bit can be coded into one symbol that is sent over the air, and we call this GPRS/GSM coding *Gaussian Modular Shift Key* (GMSK). We show GMSK coding in Figure 4.3. But as we all know, there are 360 degrees of possible phase shifts, so why not just code several bits into a single symbol? With EDGE, we use three bits per symbol in a modulation technique called 8PSK. Using three bits per symbol, one symbol can represent eight different values (two to the power of three—2^3), as seen in Figure 4.4.

In other words, every radio signal sent over the air can transport more bits of information, thus increasing the bit rate. As you can imagine, squeezing in more capacity over the air makes the reception and decoding trickier. As a result, the receiver must be more advanced, and the signal quality must be higher. This situation is a bit of a problem, however, because the signal quality for wireless

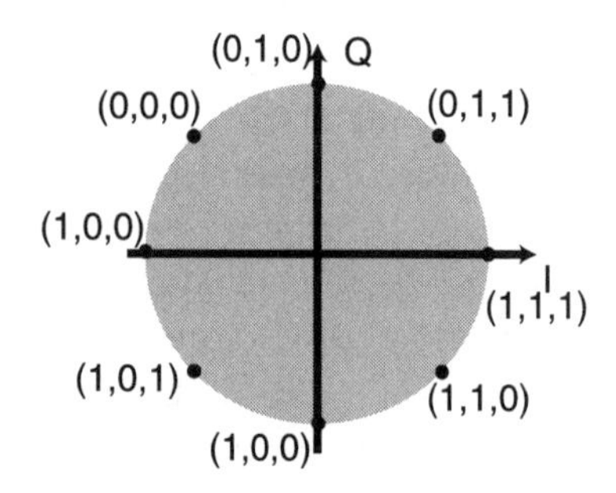

Figure 4.4 8PSK coding; three bits per symbol.

systems often varies greatly as users move around (and especially as they move farther from the base stations). To remedy this problem, EDGE provides nine different coding schemes (compared to the four that GPRS uses), and you can switch a connection between different schemes. Table 4.1 shows the different coding schemes and the resulting bit rate per time slot.

The choice of coding scheme is dynamic and depends on the current *Channel to Interference* (C/I) ratio. C/I describes how strong the received signal is relative to other signals that do not contain the desired data. As the signal quality goes down, EDGE switches to a coding scheme that is more robust but that also gives a lower throughput. In Table 4.1, the MSC-9 coding scheme requires high signal quality because it lacks most of the protective coding that MSC-1 provides.

Figure 4.5 shows how the bit rate of an EDGE user goes down as the user moves away from a base station and how the C/I dips lower. We commonly call this feature link adaptation.

The services that EDGE offers are in accordance with its legacy systems, GPRS and HSCSD. E-GPRS offers packet-data services, and E-HSCSD is the corresponding high-speed, circuit-switched service. We foresee that EDGE terminals will handle E-GPRS and E-HSCSD as well as backward compatibility with GPRS and GSM/HSCSD.

In order to facilitate the convergence between GSM and TDMA, EDGE also has access to an upgrade for TDMA networks. One obstacle here is that TDMA channels are 30Khz while GSM channels are 200Khz. An EDGE version that remedies this situation is called Compact EDGE, as opposed to the standard Classic EDGE. The compact version is for data only and uses a time-divided

Table 4.1 EDGE Coding Schemes

CODING SCHEME	BIT RATE PER TIME SLOT (KBPS)
MSC-9	59.2
MSC-8	54.4
MSC-7	44.8
MSC-6	29.6
MSC-5	22.4
MSC-4	16.8
MSC-3	14.8
MSC-2	11.2
MSC-1	8.4

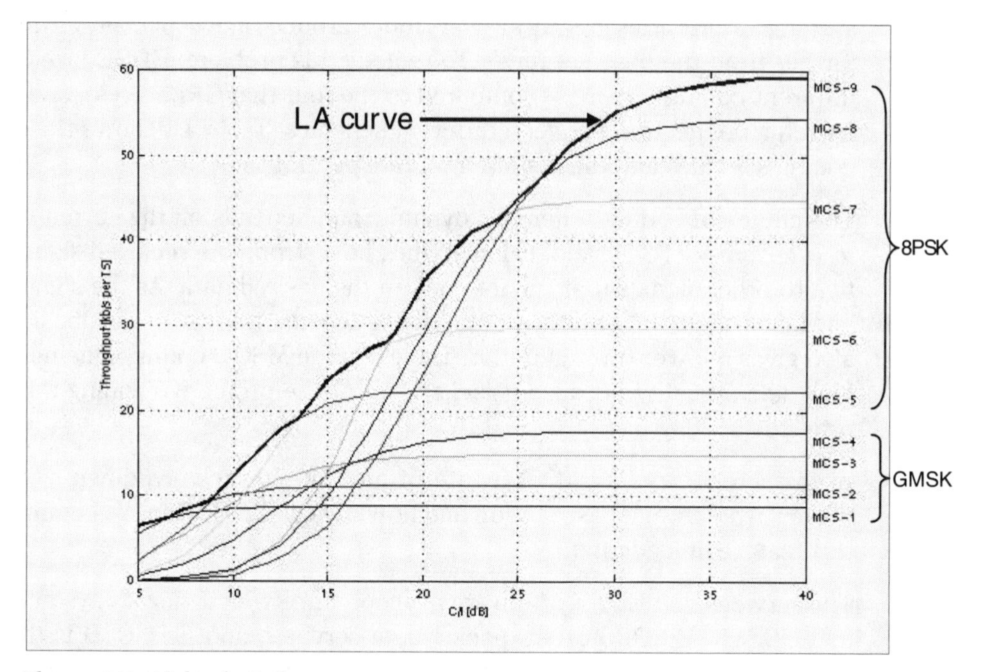

Figure 4.5 Link adaptation.

control for its 200kHz channel using less spectrum than classic (.6 MHz compared to 2.4 MHz required by classic). A TDMA operator can either choose to deploy Compact EDGE in its existing channel structure or free up some frequencies in order to implement Classic EDGE. One major benefit of this convergence is the emergence of handsets that support both systems in the not-so-distant future.

When talking about *Quality of Service* (QoS) for GPRS, we concluded that the PCU was limiting the functionality by not being capable of handling different data streams differently. We remedied this problem in EDGE, and it now supports similar QoS functionality as WCDMA does. This support is natural because GPRS, EDGE, and WCDMA core networks are defined in the same standard from 3GPP Release 1999 (the first EDGE and WCDMA release) and onward.

And then, we have the standard question, "What will the bit rates *really* be?" Well, the same reasoning as for GPRS applies here, and it mostly comes down to handset capabilities and system load. Users still share the same resources, and the 384Kbps are shared between all users on one transceiver (the same eight time slots). Likely, EDGE will at least initially deliver 30Kbps to 200Kbps in the downlink and 30Kbps to 60Kbps in the uplink.

A common question related to EDGE for GSM systems is, "Why should one choose EDGE when WCDMA comes along not too far behind?" The reasons are, of course, different for different operators, but one strategy is to gain market share among early adopters by deploying EDGE before the competition launches WCDMA. The range of applications will be more or less the same, and parts of this customer base can then be migrated to WCDMA—either by installing WCDMA radio networks or by teaming up with a partner who does. You should remember that EDGE provides substantially higher bit rates than GPRS and is a cost-efficient upgrade (even for nationwide coverage). If one operator can offer EDGE all around the country from day one, it will take time before the competing WCDMA users can catch up. In addition, if a GPRS user who needs four time slots to receive 40Kbps instead would use EDGE, he or she would only need one time slot. This situation could potentially free up capacity or at least get more revenue out of each transceiver of the base stations.

System Description: 3G CDMA Systems

The 3G *Code Division Multiple Access* (CDMA) systems specified by IMT-2000 come in three different modes, all of which serve the different needs of operators. These modes are all in the same family of 3G CDMA standards, although you sometimes will see them as two separate standards: WCDMA and cdma2000. First, we will describe some of the aspects that are common to all high-speed CDMA systems, and then we will explore some characteristics of the individual modes.

Three Modes of CDMA for 3G

Initially, there were numerous suggestions for CDMA in 3G standardization, but discussions and harmonization have now narrowed it down to one system with three different modes. All of the systems are split into a radio part and a core network part. The parts will later be interchangeable so that one radio interface can be either of the core network solutions. Figure 4.6 shows the different CDMA standards that we now call the different modes in the unified CDMA standard. In order to show a complete view of the 3G standards, EDGE is also shown, as it is also a 3G standard even though it does not use CDMA.

The result is that the following three CDMA-based radio interfaces are available (EDGE is TDMA based, like GSM/GPRS):

- DS WCDMA (called UMTS in Europe), which is intended for new spectra in the 2GHz band and uses *Direct Sequence CDMA* (DS-CDMA) in FDD

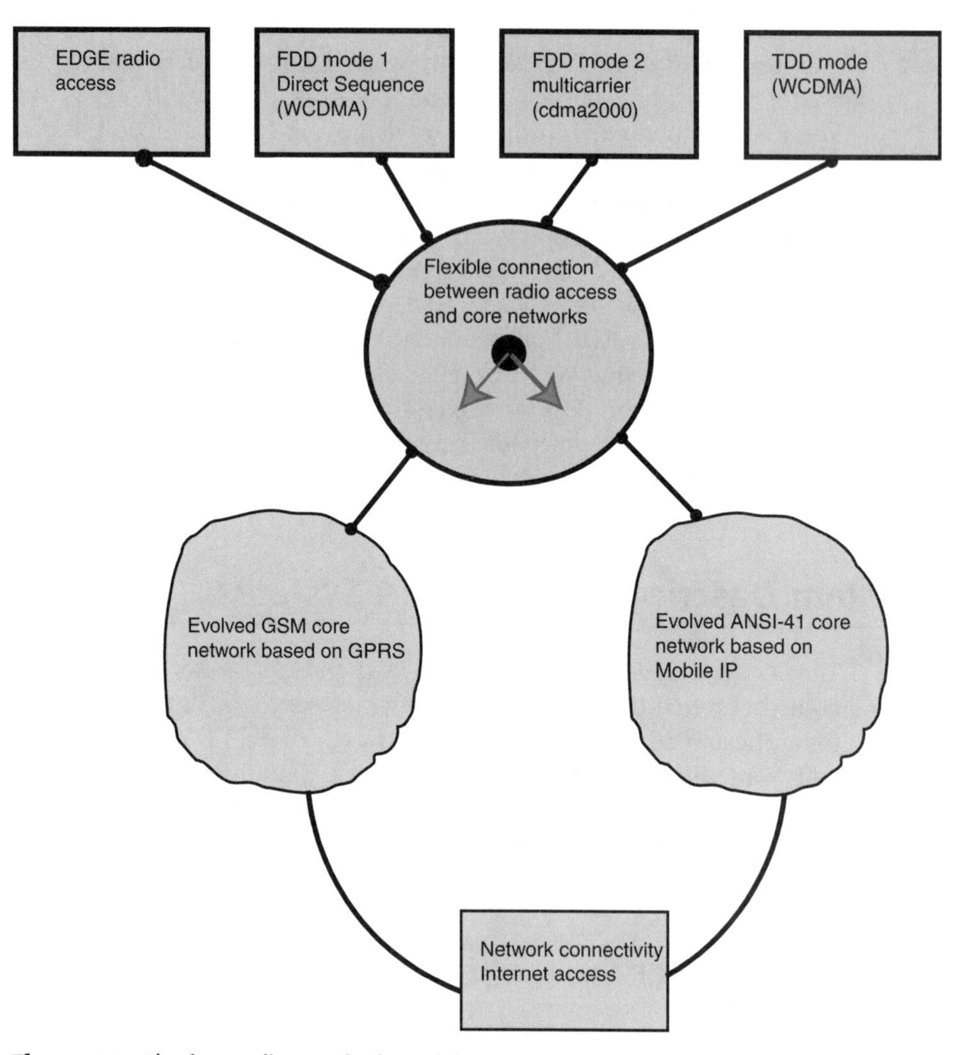

Figure 4.6 The four radio standards and the two core networks.

mode. This interface will be a common upgrade for current GSM/GPRS and PDC operators.

- WCDMA TDD mode, which is intended for new spectra in the 2GHz band and uses *Time Division Duplex CDMA* (TDD-CDMA). We foresee this system for indoor usage.
- cdma2000 MultiCarrier FDD, which does not need additional spectra but operates on the same frequencies as narrow-band cdmaOne systems. Its technology is commonly called Spectrum Overlay because it can operate in

the same spectrum as its legacy systems. Cdma2000 will be a common upgrade for cdmaOne operators.

These radio interfaces can then be connected to one of the following core networks:

- GSM/GPRS/MAP, which GSM/GPRS operators use
- ANSI-41/Mobile IP, which primarily cdmaOne and TDMA operators use

For obvious reasons, the most common configuration will initially be DS-WCDMA or TDD-CDMA with a GSM core network and cdma2000 or EDGE on ANSI-41 networks. While this view is the traditional view, more and more operators are seeing the flexibility in the standards shown in Figure 4.6. A good example is the huge U.S. carrier AT&T Wireless, which will add GPRS, EDGE, and WCDMA to its network after having been a large TDMA operator.

Features That Are Common to All 3G CDMA Systems

In order to have a new radio interface, you need new base stations and BSCs for WCDMA. But for WCDMA, a *Radio Network Controller* (RNC) replaces the BSC, and as the name indicates, it performs a wider range of tasks. While cdma2000 does not need as many hardware upgrades and can be implemented without new base stations, the following WCDMA RNC tasks can still be performed in cdma2000:

Handover. Ensuring that a mobile can move between base stations without getting interrupted. In 3G, handover is extended with the concept of diversity. In other words, a mobile can be connected to several cells at the same time and receive data from all of them. A so-called rake receiver in the mobile device collects the signals from the different sources and combines them into a better signal.

Admission control. Because users will have different kinds of subscriptions and different qualities of service, it is important for the system to prioritize users when setting up connections. The admission control function will evaluate requests for new connections and changes to existing ones in order to ensure that it makes fair decisions. If the system is fully loaded, some users might not be admitted, while high-priority users could be let in while reducing speed for others who are already connected. This process occurs in accordance with the QoS agreement that is established between the subscriber and the operator.

Code allocation. In TDMA systems such as GSM, users are separated within a cell by using different time slots (taking turns at sending). In CDMA, everyone sends at the same time, and different codes separate users. These codes are chosen so that the interference between different users and different cells is minimal. The optimal case would be if everyone could use orthogonal codes (codes that do not interfere with each other at all). This situation is possible to some extent but requires careful code planning by the system. Most of this code planning is done automatically on the network.

Power control. In CDMA, the main resource is the power—both the power that the base station uses to reach a mobile and the power that the mobile handset uses. In other words, sending farther and increasing the bit rate increases the power that is needed. The power control algorithms will ensure that a user gets data that is sent with sufficient power and that users are instructed to send at such power levels that all users arrive with the same magnitude of signal strength at the base station. Remember that everyone sends and receives at the same time in CDMA, so if someone is using a megaphone, no one else would be heard.

The base station performs roughly the same tasks as in 2G systems, although now the range of services that it must set up are increasing in number and complexity. In other words, we will build the base station somewhat differently (with pooled resources). This way, the system can set up a 128Kbps connection with the same resources that it recently used for two 64Kbps calls. In 2G systems, on the other hand, base station resources were often dedicated to one or two tasks. The antennas that are connected to the base station can be configured in a number of ways, depending on usage (see Figure 4.7). For a rural area, it can be beneficial to use an omnisector, while more traffic-dense areas are better suited for a three- or six-sector base station.

As we mentioned previously, the power of base stations and MSs is really important. As an MS moves farther away, it needs to transmit at a higher power in order to reach the base station and to keep the same bit rate. The same situation applies to the base station. Admission control, together with power control, ensures that a single MS will not consume the entire power level of one base station. In the uplink there is less flexibility, however, because an MS has a limit to how much power it can use. Usually, this problem is solved by letting the MS go down in bit rate as it moves farther away from the antenna (unless it finds a new cell that is closer to join). This factor is important to consider, because an application must be prepared to accept a lower bit rate when this situation happens and gracefully degrade service if possible (like cutting the frame rate of a video).

Because the power from the base station in the downlink is also limited, having more users in the cell will give less power to each MS. In effect, the size of the

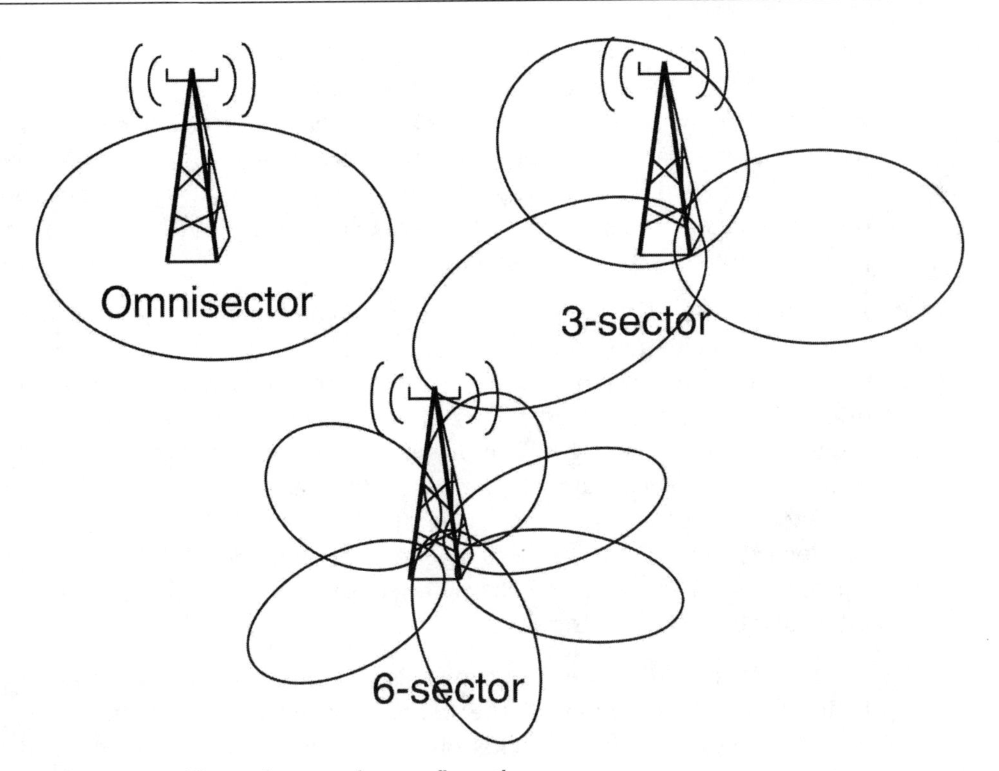

Figure 4.7 Different base station configurations.

cell becomes smaller because the base station of a fully loaded cell only has power to reach those that are closest. This phenomenon is called cell breathing and is somewhat tricky to handle (see Figure 4.8). Every infrastructure vendor must tackle this issue, and the Admission control function can help.

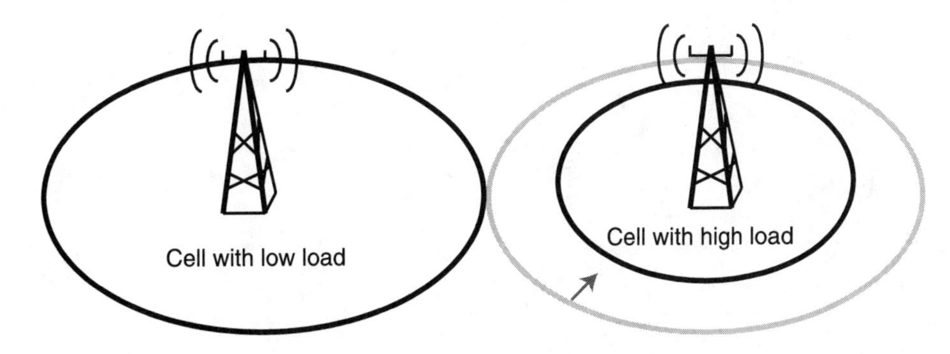

Figure 4.8 Cell breathing.

Cell breathing also becomes less of an issue with the introduction of hierarchical cells. A large cell can cover the same area as a number of smaller cells, acting as the backup for the smaller ones and as the preferred cell choice for users who are moving fast. These larger cells are often called macro cells, and we call the smaller ones micro cells. In Figure 4.9, the car should preferably be connected to the macro cell, or it would have to change micro cells very often, thus putting a larger load on the system.

We predict that 3G terminals will be capable of performing many different tasks at the same time, so it is essential to add this functionality in a way that is efficient for both the network and for the terminal. The intuitive way of handling this situation is to add more physical channels to one device as the user requests additional services. The problem, however, is that the terminal needs to have transceiver equipment for each physical channel, which adds lots of costs to the handsets. To remedy this situation, 3G CDMA systems enable several logical channels (voice, data, packet switched, and circuit switched) for a single, physical channel. The characteristics of each channel are chosen separately and can be very different.

We do not predict that the core networks will affect the end user and the application to the same extent as the radio interface will. Most developers have already gotten used to the packet data core networks, as we described in the previous chapter.

WCDMA-Specific Features

There are really two different modes within the WCDMA standard: *Frequency Division Duplex* (FDD) and *Time Division Duplex* (TDD). TDD mode operates in the unpaired band at 2010–2025MHz and is a mix of TDMA and CDMA.

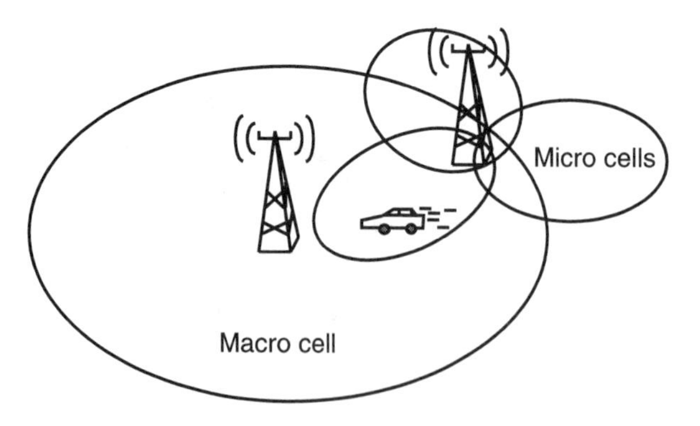

Figure 4.9 Macro micro cells.

This mode is mostly used indoors and is likely to be available a bit later than DS-CDMA. Therefore, we will not describe it in further detail here; rather, we will focus on FDD mode.

As mentioned previously, the WCDMA standard is developed by 3GPP, and released roughly on a yearly basis. The traditional naming has therefore been 3GPP Release '99 (the first WCDMA standard), Release '00, and so on. You might also encounter the use of Release 3 and Release 4, which are synonymous with Release '99 and Release '00 respectively. This is also the way that the standards documents are numbered.

As we mentioned previously, WCDMA introduces a new radio interface that almost always is implemented in a new spectrum. The primary spectrum that is used for DS-WCDMA (from now on, only called WCDMA) is 1920–1980MHz for the uplink and 2110–2170MHz for the downlink. This band is mostly just described as the 2GHz band (depicted in Figure 4.2).

3GPP Release 1999 describes the first release of WCDMA, and this release offers both packet data and circuit-switched services. For each kind of service, a number of bit rates are available for the channels. For the packet data channels, the system changes between these different bit rates without the user noticing—maximizing the total capacity of the system. You can calculate the bit rate for a channel as the chip rate divided by the spreading factor, as seen in Table 4.2.

The spreading factor determines how large a code you will use when spreading the data. The chip rate is system specific and indicates the rate of bits that are sent over the air. For WCDMA, the chip rate is 3.84Mcps, or 3,840 bits per second. (A bit over the air is commonly called a chip.) A significant change from the GPRS that we used for 2.5G and the upgrade for 3G is the improved QoS support. Now, the QoS is not only supported in the core network, but also all

Table 4.2 Spreading Factors for Different Bit Rates

USER BIT RATE (KBPS)	SPREADING FACTOR	CHIP RATE (MCPS)
30	128	3.84
60	64	3.84
120	32	3.84
240	16	3.84
480	8	3.84
960	4	3.84
1920	2	3.84

the way to the user. The core network part is common for GPRS, EDGE, and WCDMA, because these systems will have a common core network from 3G and onward. The QoS profile for a connection is negotiated at the PDP context update (usually when the device is turned on). Both the handset and the network then know this profile. As a connection is requested from the handset, this profile is then used to configure the resources that are needed in the radio part and on the core network. The following QoS classes are available (see Table 4.3):

Conversational class. This class is ideal for real-time applications where it is essential for delay to be minimized. The traffic for such services is usually symmetric or close to it. Those applications that need packets to arrive in a stream where the distance between packets is kept constant will use this class.

Streaming class. In the streaming class, the focus is on delivering data in a steady stream (in other words, keeping the distance between packets constant). The emphasis is not so much on the delay, because most applications of this class are expected to be asymmetric with fewer interactions than the Conversational class. In typical streaming applications (video, audio, text tickers, and so on), the information can start to show on the client device before the entire file has been received.

Interactive class. Here, there is a constant exchange of information between the user and the network, but it is not as time critical as in the Conversational class. This exchange can be a regular information search with a browser,

Table 4.3 QoS Classes

TRAFFIC CLASS	CONVERSATIONAL CLASS	STREAMING CLASS
Characteristics	▪ Low delay ▪ Preserves time relation between packets	▪ Preserves time relation between packets
Typical applications	Voice, highly intensive games	Streaming media
TRAFFIC CLASS	**INTERACTIVE CLASS**	**BACKGROUND CLASS**
Characteristics	▪ Request response pattern ▪ Preserves data integrity	▪ Best effort ▪ Non-time critical data ▪ Preserves data integrity
Typical applications	Web browsing	Background synchronization, downloads, and so on

chat applications, or location-based applications. Typically, the user requests some information and a server on the network side responds with the appropriate information. Low-intensity games are also expected in this class because it will most likely be cheaper to use than the more demanding Conversational class.

Background class. For those applications that are not time critical at all, the Background class is appropriate. This class includes tasks that run in the background or perhaps while the user is not actively using the device. The delay can be seconds or even minutes depending on the network load, but the cost is likely to be lower than the other QoS classes. This class makes it appropriate when large chunks of data need to be exchanged, as in the synchronization of calendar and e-mail data, or when a new version of an application is to be downloaded.

The QoS can, for instance, be set in the MS when the connection is established, just like we described in the previous chapter when the PDP context was set in order to reflect the desired profile. Some examples of parameters include the following:

Traffic class ("conversational," "streaming," "interactive," and **"background").** To choose the desired QoS class.

Maximum bit rate (Kbps). In order to declare the chosen channel (used to reserve how much capacity the system hands you).

Guaranteed bit rate (Kbps). Can be used in order to make sure that the application gets a certain bit rate throughout the session.

Transfer delay (ms). Specifies the maximum delay for 95 percent of the packets.

We describe the QoS aspects of WCDMA in detail in 3GPP TR 23.907 and TS 23.107, where these parameters and all of the others are specified.

In order to facilitate end-to-end QoS, most vendors enable mapping of the wireless QoS profile that is negotiated at PDP Context Activation to the QoS mechanisms of the concerned networks. This process might include mapping PDP context to DSCP values in DiffServ networks to MPLS labels in MPLS networks to ATM Service Types and to Frame Relay Traffic Parameters, and so on.

How such mapping is done is a matter of operator configuration. Given the scarcity of radio resources, however, data to and from wireless networks should be given a relatively higher priority compared to the same type of data when transmitted purely over fixed networks. With this approach, where no admission control for the fixed network is performed at the time of the context activation, careful network design must ensure that adequate resources are built into the fixed networks.

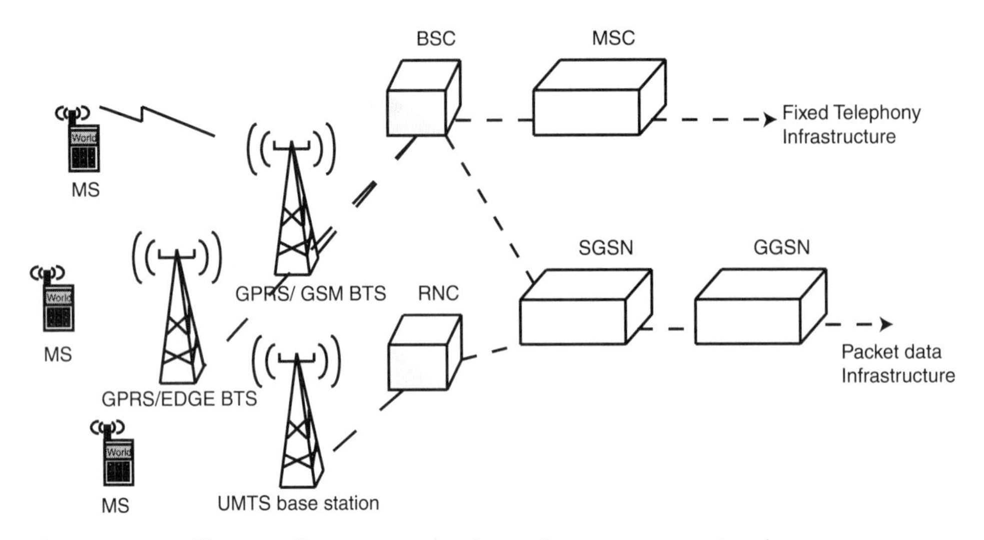

Figure 4.10 Different radio access networks on the same core network.

Figure 4.10 shows how the base stations and RNCs of a WCDMA radio network are attached to a regular GSM/GPRS core network.

Note how the GSM/GPRS/EDGE base stations are kept and used even as WCDMA is introduced. Some solutions even make it possible to insert a WCDMA transceiver into an existing GSM base station. Radio access then attaches to the existing circuit-switched (GSM) and packet-switched (GPRS) core networks. In the figure, we make several simplifications in order to make it easier to understand the basic thinking. The handsets can then talk to either base station that supports the technologies that it supports. Seamless handovers between WCDMA and GSM/GPRS are included in the first release of the standard, and even the first batch of handsets are expected to support this functionality by using dual-mode GSM/GPRS and WCDMA. In other words, GSM/GPRS can be used for rural areas as WCDMA is being built, because WCDMA requires coverage to be built from scratch once again. We also predict support for handsets that also handle EDGE (in addition to WCDMA, GSM, and GPRS), although at the time of this writing, we do not know when and to what extent.

Features That Are Specific to cdma2000

The cdma2000 air interface is introduced in two phases: 1X and 1xEV, both of which are backward compatible with legacy cdmaOne systems. The air interface is similar to WCDMA with one major difference: It is developed in 1.25 MHz of spectrum, so it is therefore ideal for deployment in an existing spectrum on top of existing cdmaOne networks. Cdma2000 1X will be implemented during 2001 by most cdmaOne operators, and has already been implemented in Korea. The upgrade is simple and similar to the GPRS upgrade for GSM operators. The functionality is mostly added with new software, and it can be done quite quickly. While cdma2000 1X can deliver theoretical bit rates of more than 200Kbps, the figure 144Kbps as an average bit rate is commonly used.

While the cdma2000 1X standard now seems stable and commonly accepted by everyone, the migration after that was still an open issue for many operators in 2001. Within the standardization group, there were three proposed migration steps for 3G (note that some call cdma2000 1X a 3G system):

- 3X RTT
- 1eXtreme
- 1xHDR

In addition, a system based on WCDMA radio and an ANSI-41 core network was being considered, for those who could get access to the spectrum.

Thanks to efforts within the CDMA community, the evolution for CDMA systems beyond cdma2000 1X is now laid out. The evolution of cdma2000 beyond 1X is now labeled cdma2000 1XEV. 1XEV will be divided into two steps: 1XEV-DO and 1XEV-DV, where 1XEV-DO stands for 1X Evolution Data Only and 1XEV-DV stands for 1X Evolution Data and Voice. Both 1XEV evolution steps provide for advanced services in cdma2000 by using a standard 1.25 MHz carrier. Evolution with cdma2000 will therefore continue to be backward compatible with today's networks and forward compatible with each evolution option.

1XEV-DO will be available for cdma2000 operators during 2002 and will provide for higher data rates on 1X systems. 1XEV-DO will require a separate carrier for data, but this carrier will be capable of handing off to a 1X carrier if simultaneous voice and data services are needed. By allocating a separate carrier for data, operators will be able to deliver peak rates in excess of 2Mbit/sec (best effort) to their data customers.

1XEV-DV solutions will be available approximately one and a half to two years after 1XEV-DO. IXEV-DV will bring data and voice services for cdma2000 back into one carrier. A 1XEV-DV carrier will provide not only high-speed data and voice simultaneously but also will be capable of delivering real-time packet services. This functionality means that QoS for higher data rates beyond 144 kbps will be introduced in cdma2000 networks with the advent of 1XEV-DV.

The Mobile IP/Simple IP cdma2000 Core Network

cdma2000 introduces a new set of network elements—the *Packet Core Network* (PCN)—in order to support packet data services. The PCN consists of the following elements. An example cdma2000 solution includes a *Packet Data Serving Node* (PDSN), a *Home Agent* (HA), and an *Authentication, Authorization, and Accounting* (AAA) server.

The PDSN functions as a connection point for the radio network and for the ATM/IP network. The PDSN also facilitates the AAA activities with its connection to both the HA and the AAA server. In addition, it provides a foreign agent function, where it registers and facilitates services for network visitors. You can see this setup in Figure 4.11, where a user visits a network and becomes connected to the foreign agent that then communicates with the user's HA.

In conjunction with the PDSN, the HA authenticates Mobile IP registrations from the mobile client and maintains current location information for the mobile. The HA also performs packet tunneling; that is, the HA receives packets that are destined for the mobile client's permanent address and routes them to the mobile's new temporary address. The AAA server authenticates and authorizes the mobile client, providing network security, QoS, and the storage of user accounting information that is received from the PDSN. This architecture is extremely flexible, and as the user is located at different places on the network, different foreign agents are used.

In summary, the nodes in the cdma2000 core network have the following roles:

Home Agent (HA). Authenticates Mobile IP registrations from the mobile station and redirects packets to the foreign agent in question. The HA can

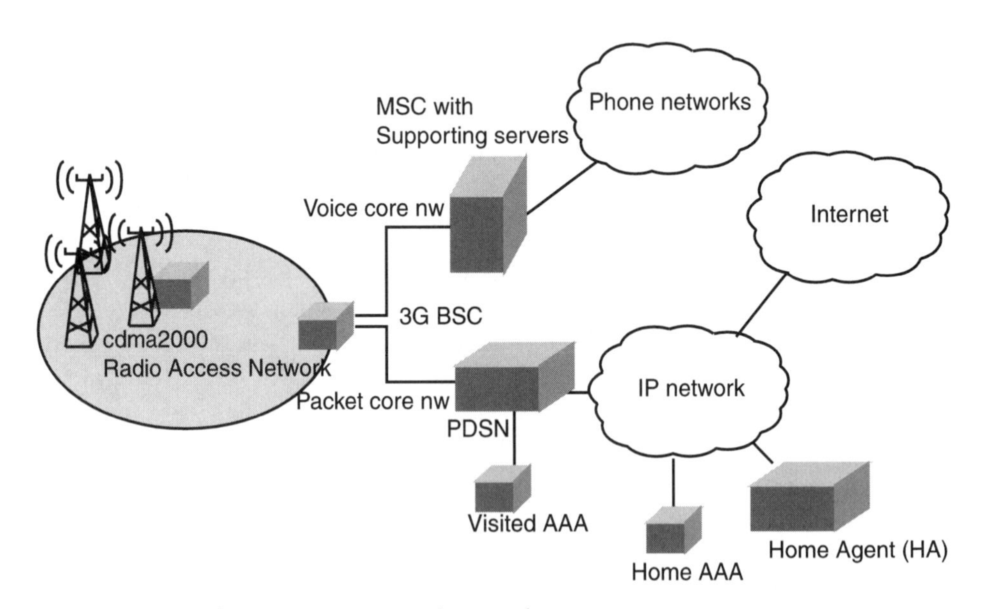

Figure 4.11 A cdma2000 core network example.

also optionally provide secure communications to the PDSN and assign a dynamic home address. The HA also receives provisioning information from the AAA for users.

Packet Data Serving Node (PDSN). Establishes, maintains, and terminates *Point-to-Point Protocol* (PPP) sessions at the mobile station. When simple IP is used, the PDSN also assigns IP addresses to the users. At this point is where the foreign agent is commonly placed, to which visiting users are connected. The PDSN also initiates the security sessions that the AAA provides.

Authentication, Authorization, and Accounting (AAA) server. The AAA is responsible for most of the security aspects of the core network in cdma2000. The AAA can authorize users and provide user profiles and QoS information to requesting nodes and can also optionally assign IP addresses.

There are two basic configurations of the cdma2000 core network:

Simple IP. Simple IP refers to the access method in which the user is assigned a dynamic IP address from a service access provider. The user can keep the IP address within a certain network-dependent (geographical) area. If the user moves outside this area, the user cannot maintain the IP address. This model is the simpler model (with restricted mobility).

Mobile IP. Mobile IP refers to the access method that is based on the Internet standard that is specified in RFC 2002. Here, the user can use either a static or dynamic IP address belonging to its home IP network. The mobility options here seem endless because the user can maintain his or her IP address no matter whether it moves throughout the cdma2000 network or throughout other networks. The home agent then resides in the cdma2000 network while the foreign agent is close to the user's current position.

The main advantages with Mobile IP as a core network solution are that it is based on an existing *Internet Engineering Task Force* (IETF) standard, which makes its components cheap, mass-market routers, and the long-term vision of seamless IP access. Seamless IP access here means that you could have a laptop with both a *Wireless Local Area Network* (WLAN) and cdma2000 radio cards and keep the same IP address as you move out of WLAN coverage into cdma2000 coverage. A similar all-IP approach will also appear in WCDMA but not in the first release.

Which 3G System Is the Best?

As always, when there are multiple choices, people in general (and media in particular) want to see a battle where one tries to find out who is the best and who will win. The 3G evolution is no different, and with lots of money at stake, there has been a substantial effort by the parties who are involved to promote individual technologies. The problem that is associated with this kind of technology fight is the massive amount of nonobjective information. If you search the Internet for information about 3G technologies, you will likely find that most of the information is biased toward one technology or the other. We are not saying that these sources are lying, but there are always ways of showing off things that are to the advantage of either technology. The maximum bit rate and coverage are often shown without stating whether there are one or 100 users in that cell—a small detail that can change the result by 90 percent (say that a GPRS cell with two heavy, eight-time-slot FTP users are competing with you).

The optimal choice of 3G technology for an operator depends on the following parameters:

- Do you have a license for the 2GHz band?
- What system are you using now for 2G/2.5G?
- What systems are your competitors using?
- What systems exist in the neighborhood in which your subscribers are likely to roam?
- How big is the area that needs to be covered, and what is the cost of covering it?
- How much do you want to spend?

Thus, the optimal choice of 3G technology is to carefully choose and evaluate the individual needs of the operator, regarding both business and technology, and from there decide where to go next.

Future Applications and Enhancing Applications for 3G?

While 2.5G and wireless packet data was a big step for many application developers (many of whom came from the fixed Internet world or from circuit-switched mobile networks), the step might actually be smaller when you are moving to 3G. As we will see in later chapters, there are two main tracks to follow when approaching a new technology: overcome the challenges and leverage the possibilities. With 3G, it is easy for most people to see the advantages of higher bit rates, better QoS, and multiple, simultaneous services. Then, it is easy to get excited about all of these features and overlook the challenges that actually arise with the advent of such a complex system. How do you make the *Man-Machine Interface* (MMI) usable when there are dozens of available circuit-switched channels, different packet data channels, and four QoS classes-each having tens of adjustable parameters? Although many of these issues cause headaches for device manufacturers, the application developers are still affected. Broad adoption of new features and technologies has always been highly dependent on making the features easy for the user. Most people would be hesitant about purchasing a new gadget if it will make their life more complicated, although it has some nice features. The challenge here is to take all of these new opportunities and turn them into a rich (but still user-friendly) experience.

Because most of the 3G terminals will support legacy systems in some way, you must take that factor into account. Say that you have developed a very nice streaming media application for WCDMA and are running it in Sweden together with an operator. Sweden has a very low population density in its northern parts, and network buildout has always been slower in that area. In other words, initial WCDMA coverage is limited to larger cities and freeways in the north (although the companies in the Swedish UMTS license bid are making hefty promises). So, what happens to this streaming application when a user moves off the freeway and into areas that only have GPRS coverage? Will the frame rate go down, or will the quality degrade? Maybe the moving pictures are removed and only sound is played. The big issue is not *what* to degrade but rather knowing *why*, because it all depends on the use of the application. A news feed is naturally stripped of the moving pictures as the bit rate goes down, but a doctor who is monitoring patients might prefer removing the colors or dropping the sound. Most of the time, this situation is not a hard thing to change, but the work lies in understanding the needs of the users and enabling the graceful degradation of service.

More than likely, you will find the new possibilities of having multiple services that each have an individual QoS profile really interesting. You could potentially have one part of an application running at a high QoS, delivering the critical updates, while the rest is handled by a lower QoS and (hopefully) by a subsequently lower cost. In a real-time game, the information that updates the characters on the screen could be given a higher priority than chat messages between players, for instance.

Summary

There are several 3G systems, all of which are created with migration from legacy systems but with a different degree of backward compatibility. The EDGE and cdma2000 radio interfaces can be installed in existing spectrums, while new frequencies are usually allocated for WCDMA. Although the systems are different, there are many common features (such as higher bit rates, more services, and QoS). The 3G systems will start to appear during 2001 in Japan and a bit later (2002) in the rest of the world.

CHAPTER 5

Bluetooth—Cutting the Cord!

The cellular systems that we described in previous chapters bring content and information from all around the world to our devices. As a result, we can access content that resides on a remote server or perhaps on another user's mobile device. Developers designed and optimized these systems for this purpose, and these devices enable a high degree of mobility and range. If you are riding in someone's car and want to be updated with the results of an ongoing hockey game or want to view a chart of today's development of a stock that you are following, a *General Packet Radio Services* (GPRS) or *third-generation* (3G) phone will do the job. Then, perhaps you want to share the video clip or stock chart with the other people in the car. Using the cellular network to send the data via all of the network nodes and then back out again (probably through the same base station) is not logical, however. In this case, a short-range radio technology would come in handy. This example illustrates just one of the many applications that Bluetooth will facilitate. Bluetooth not only eliminates the need for many of the cables that we use today, but it also enables us to extend the usage of the information that the cellular systems bring to us. Bluetooth is a short-range radio technology that is complementary to other wireless technologies, and we expect Bluetooth to reside in hundreds of millions of devices in the future due to its small size and relatively low price. For the application developer, Bluetooth represents a great opportunity for ubiquitous computing that will foster a huge variety of exciting products.

Background and History

Bluetooth originated as a means for cable replacement, because computer and cellular telephone users commonly view this process as a hassle. The resulting radio technology does much more than that, however, and provides a key enabler for ubiquitous computing—connecting just about every device in our surroundings.

In 1994, Ericsson Mobile Communications initiated a study in the southern parts of Sweden in order to find a low-power and low-cost radio technology for cable replacement. Mobile hands-free devices and other accessories were somewhat limited in that they ironically needed a wire to connect to the wireless phone, and this type of radio technology could remedy the problem. The requirements included the following:

Low power. In order to install an application in just about any device (down to cellular phones and mobile hands-free devices), the power drain from the radio chip had to be close to zero.

Low cost. In order to make most consumer electronics devices (and others, of course) Bluetooth-enabled, the cost had to be very low ($5 to $10).

Small footprint. Again, the small devices that developers envisioned as targets for Bluetooth would not permit a large chip (that would compromise their size).

Speech and data transmission. The technology had to enable both speech and data transmission, preferably at the same time.

Worldwide capability. It all had to work around the world.

After the pre-study, researchers concluded that these requirements could be met, and Ericsson also understood early on that one company alone could not achieve this goal. Other companies had, of course, had thoughts in the same direction—and keeping things proprietary would clearly have made broad adoption slower. Many times before, having several competing technologies in one area slowed growth. In order to gain wide industry acceptance, players from diverse industries and continents gathered together. All of the involved parties saw the need for using the expertise from all of the involved parties in order to avoid a market situation of many competing, incompatible standards.

In 1998, Ericsson, Intel, IBM, Toshiba, and Nokia formed the Bluetooth *Special Interest Group* (SIG). The group represented the Americas, Europe, and Asia, as well as all of the needed industry segments. Work started in a standardization environment in order to create a de facto standard for interoperable short-range radio interfaces and software. In late 1999, another milestone occurred when a number of additional heavyweight players joined the SIG. Today, the

Bluetooth SIG includes promoter companies 3Com, Ericsson, IBM, Intel, Lucent, Microsoft, Motorola, Nokia, and Toshiba, and more than 2000 Adopter/Associate member companies. The Bluetooth SIG Web site is the official source for specifications and updates: www.bluetooth.com. The name Bluetooth comes from a Danish king named Harald Blåtand (Bluetooth in English), because he symbolized unity between different groups of people.

Main Features

Of all of the requirements of the initial pre-study, only the last one became somewhat compromised. Not all of the frequencies allocated work in all countries (we will talk more about that topic in the Air Interface subchapter). The study's results met all of the other requirements: low power, small size, and low cost (although the initial chips, for natural reasons, were more expensive than the desired target cost). As volumes rise, however, researchers expect this scenario to change—and Bluetooth chips can be deployed in just about every electronic device without a significant increase in cost.

Bluetooth radio technology enables any device with a chip to communicate seamlessly, even if there are non-metallic walls or other objects in the way. In other words, you can have a laptop on your lap when you are riding on a bus while your 3G phone lies in your bag or your pocket, and you can still check your e-mail with the laptop. The laptop talks to the phone via a Bluetooth connection to the phone, which in turn connects to the mail server via a 3G wireless network (see Figure 5.1). In a similar way, the cell phone can use a wireless headset, and you can use the same headset together with a *Personal Digital Assistant* (PDA) or laptop (regardless of the brand). In order to facilitate this interoperability, developers have defined a number of profiles and usage models—and we will discuss these items in the Bluetooth Profiles and

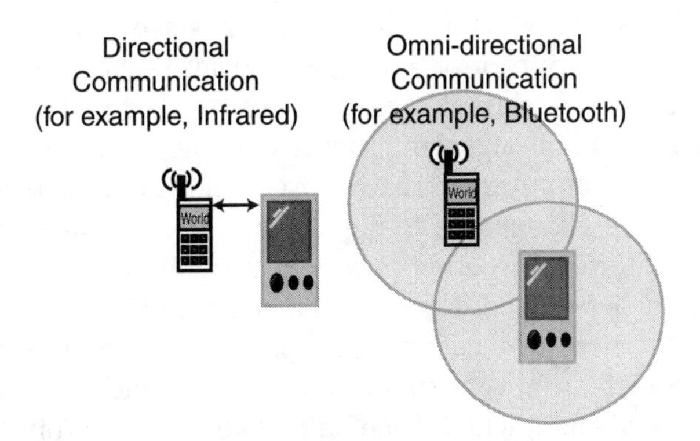

Figure 5.1 Omnidirectional versus directional communication.

Use cases subchapter. Bluetooth communication is, however, not limited to point-to-point links, and small LANs can be set up. At a meeting, the participants can create a so-called piconet and share documents. Similarly, a Bluetooth device can serve as a short-range Internet bridge that enables others to connect to it and to reach Internet content.

Short-range radio in this context means 10m, although it is possible to get as far as 100m when using a higher power. The radio signals are omnidirectional, which means that a Bluetooth signal propagates equally in all directions (see Figure 5.1)—and you do not have to point the devices at each other. While this feature means that Bluetooth is less sensitive to how the transmitting devices are aligned, it also opens up the risks of eavesdropping. To address this concern, developers created an elaborate security architecture. The maximum gross bit rate is 1Mbps, although protocol overhead limits the net throughput to 722Kbps for asynchronous transfer and 433Kbps for symmetric transfer. As with most wireless systems (such as GPRS and 3G), the bit rate also varies with the amount of protective coding (723Kbps without protective coding).

Bluetooth Air Interface

The Bluetooth air interface might look complicated at first glance (with its anti-interference frequency-hopping scheme), but a closer look shows that it is not too difficult.

Frequency-Hopping Radio

Bluetooth operates in the 2.4GHz *Industrial-Scientific-Medical* (ISM) frequency band (or 2402Hz–2480Hz to be exact). Each channel is 1Mhz wide, so there are 79 different channels. The 2.4GHz band is unlicensed, which means that anyone can operate in that location. Bluetooth devices will coexist in the same frequency as *Wireless LANs* (WLANs) and microwave ovens; consequently, the band has to be very robust. Spread-spectrum technologies help avoid interference between radio technologies. WCDMA uses a direct sequence technology where codes spread the signal in order to occupy a larger portion of the spectrum. Bluetooth uses another alternative: frequency hopping (see Figure 5.2). A Bluetooth device changes its frequency in a pseudo-random way 1,600 times per second. Interference often occurs in a small portion of a frequency band, so hopping between different frequencies makes the channel insensitive. If one packet becomes corrupt, the packet is later re-sent on another frequency on which it is very unlikely that the same interference exists. Packets are also very small, which benefits everyone from a robustness point of view.

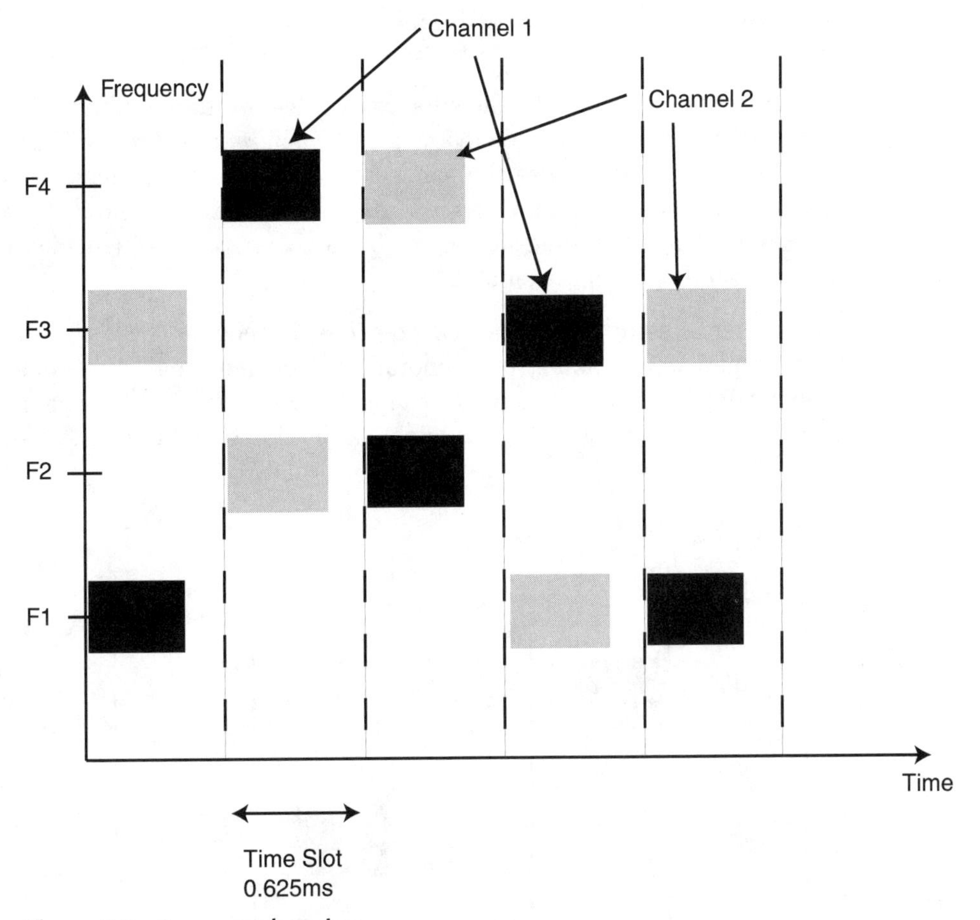

Figure 5.2 Frequency hopping.

In Figure 5.2, we show only four of the available frequencies and only two channels. Channel 1 hops from F1 to F4 and then from F2 and F3 on to F1 again. If a static disturbance (occupying one frequency all of the time) appears on F2, only one of these packets will potentially be lost (provided that the disturbance is powerful enough to corrupt that packet).

A Bluetooth channel always consists of a master and one or more slaves—the master often being the one that initiated the connection. The master decides on a hopping scheme that is related to its internal clock. The slave(s) calculates an offset, which is the difference between the master and slave clocks, and uses this information to determine the frequency to which it will hop. This process enables the master and its slaves to hop to the same frequencies at all times.

The uplink and downlink channels for one device are time multiplexed (separated in time as opposed to frequency) and *Time Division Duplex* (TDD); consequently, both channels use the same frequency-hopping scheme. In Figure 5.3, we show the same channels as in Figure 5.2, but here we show the sending parties of Channel 2. In time slots 1 and 3, the master is sending while the two slaves are sending in the other slots.

When a user is using more than two devices, it becomes hard to talk about uplinking and downlinking, but generally, all devices that share one 1MHz channel will send and receive on the same channel. Note that one channel here means that the 1MHz channel is available at all times (although the frequency

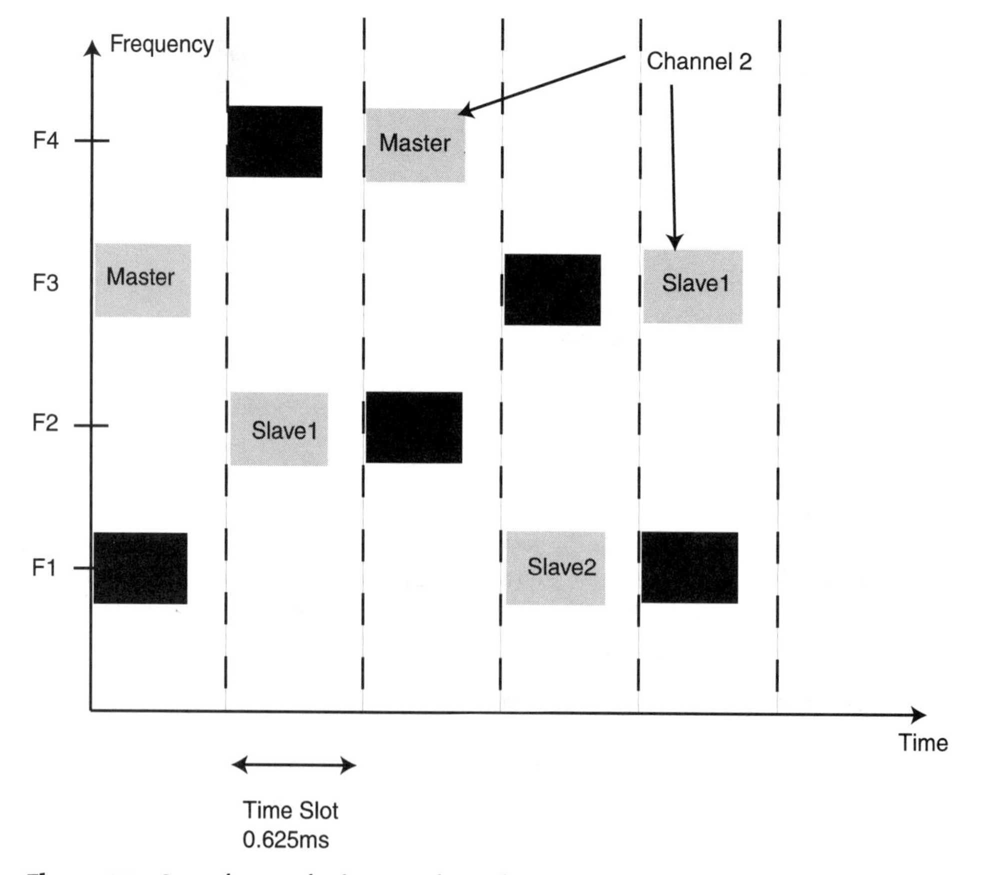

Figure 5.3 Several users sharing one channel.

itself changes every 0.625ms). So, in a piconet of eight users, all of them (uplink and downlink) have to share the available 1MHz—not leaving much capacity per user. This issue can be overcome, however, and we will discuss this topic later in this chapter.

Link Types

In order to facilitate the efficient delivery of both time-sensitive voice traffic and bursty data traffic, two physical link types have been defined:

***Synchronous Connection-Oriented* (SCO) link.** This link type is especially suitable for circuit-switched services where low delay and high QoS is required. The channels offered are symmetric (the same speed in both directions) and synchronous (both parties know exactly when the next packet will come), which is achieved by reserving two consecutive time slots at fixed intervals. Voice primarily uses this kind of link, but data support is also available on the 64Kbps channels. The advantages of sending data over a synchronous link are the same as for voice—low delay and high QoS. In addition, sending over a synchronous link produces less overhead because no header information is needed. The disadvantages are the lack of flexibility (fixed bit rate) and the waste of radio resources, because the transmission slots are constantly allocated throughout the session.

***Asynchronous Connection-Less* (ACL) link.** For data transfer and other asynchronous services, the ACL link is more efficient. This link offers packet switching, and transmission slots are not reserved but rather are granted by a polling access scheme. When two Bluetooth units are communicating, a piconet is established. A piconet is a collection of up to eight Bluetooth units where one is a master unit that controls the transmission and hopping scheme. The master indicates to a slave that it wants to send, and the slave then receives. The slaves can then send on slots only when they are in agreement with the master.

One connection can then contain several links of either type. For example, two phones can maintain a voice conversation while simultaneously exchanging data (business cards, pictures, and so on), as shown in Figure 5.4. By using a piconet, one master could maintain various SCO and ACL connections with several slaves. There is, however, a three-voice call limit within a piconet.

Because voice is highly delay-sensitive and it is not useful to have lost packets retransmitted, SCO is used for voice traffic. In other words, the voice packets

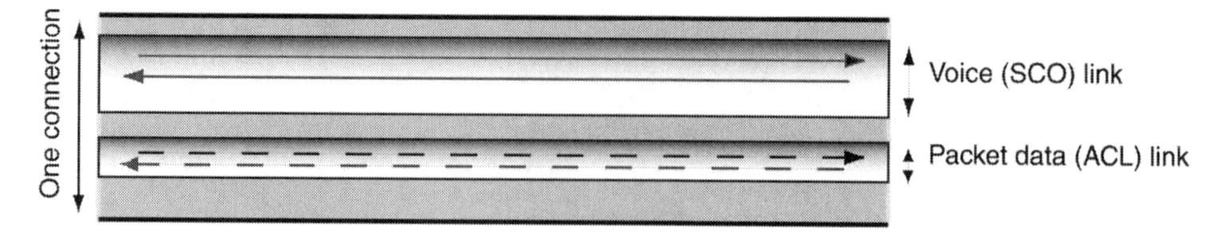

Figure 5.4 ACL and SCO links sharing one connection.

do not have to compete with other traffic concerning the available capacity, and an entire 64Kbps channel is allocated.

There are two ways to code the analog voice information into digital signals in Bluetooth. The first method is the traditional *Pulse Code Modulation* (PCM) technique, which many land-line phone networks use. PCM samples the waveform 8,000 times per second and then encodes each sample into an eight-bit sequence. This process gives 64,000 bits per second (8,000 × 8) in order to transmit the digital version of the voice information. We illustrate this process in Figure 5.5.

The second modulation technology is called *Continuously Variable Slope Delta* (CSVD) and is more immune to interference; therefore, it is more fit for the wireless environment. With CSVD, the samples are only one bit long as opposed to the eight bits that PCM uses. Therefore, the amount of information per sample is much lower, and one sample becoming corrupt will not matter all

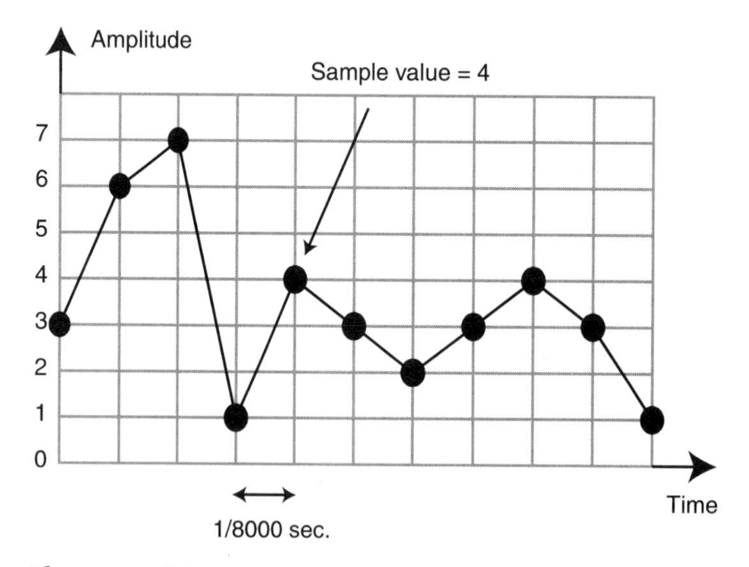

Figure 5.5 PCM.

that much (because of the increased robustness). The bit that is sent indicates whether the input signal is increasing or decreasing (in other words, the slope change of the signal is communicated). While this view is a simplified view of CSVD, it is enough to gain an understanding of it. The choice of desired modulation is made by the Link Managers of each device, but we expect CSVD to be more commonly used. Figure 5.6 shows how each part of the analog waveform is converted into a digital bit stream by using CSVD. The bits indicate the change in amplitude slope that is necessary in order for the reference signal to mimic the input signal.

In Bluetooth, voice calls are coded into 64Kbps bit streams (compared to 8Kbps for a 2G mobile phone), resulting in a significantly higher voice quality.

Although voice is always called a killer application, it was also crucial to get good packet data support into Bluetooth. Packets can either be sent one slot at a time or by using multi-slot. With multi-slot, a large packet is sent over several slots and the frequency remains the same while sending. Figure 5.7 shows two channels (as in Figure 5.3), but here the master starts by sending a multi-slot packet over three slots. In addition, Slave1 then sends a two-slot packet. Note

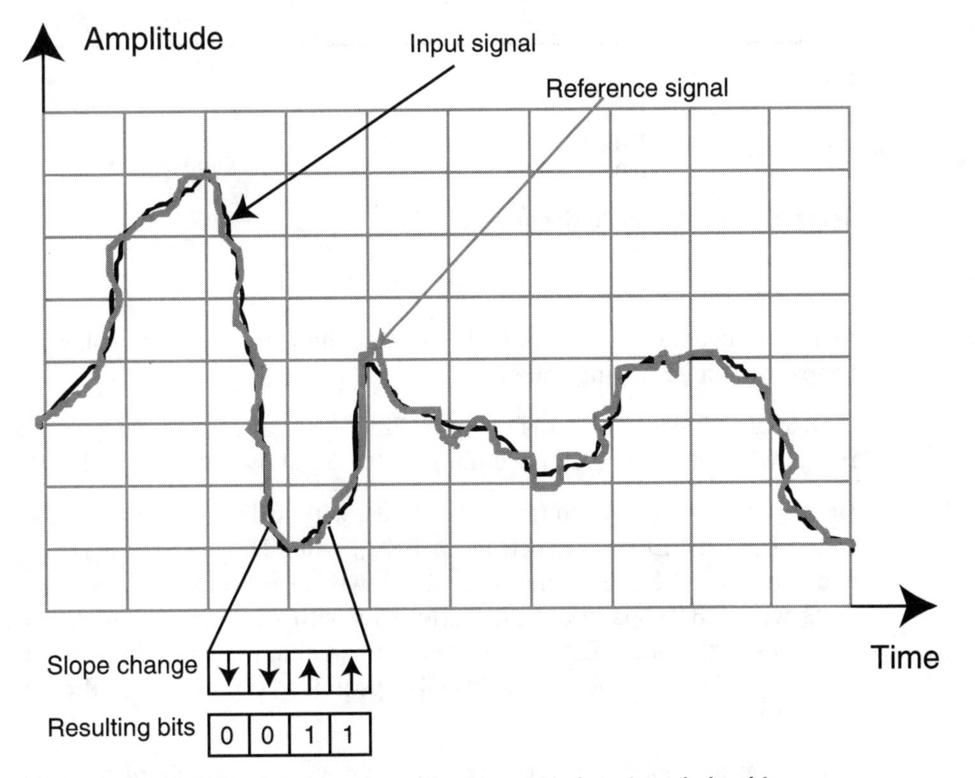

Figure 5.6 CSVD modulation where the change in slope is coded as bits.

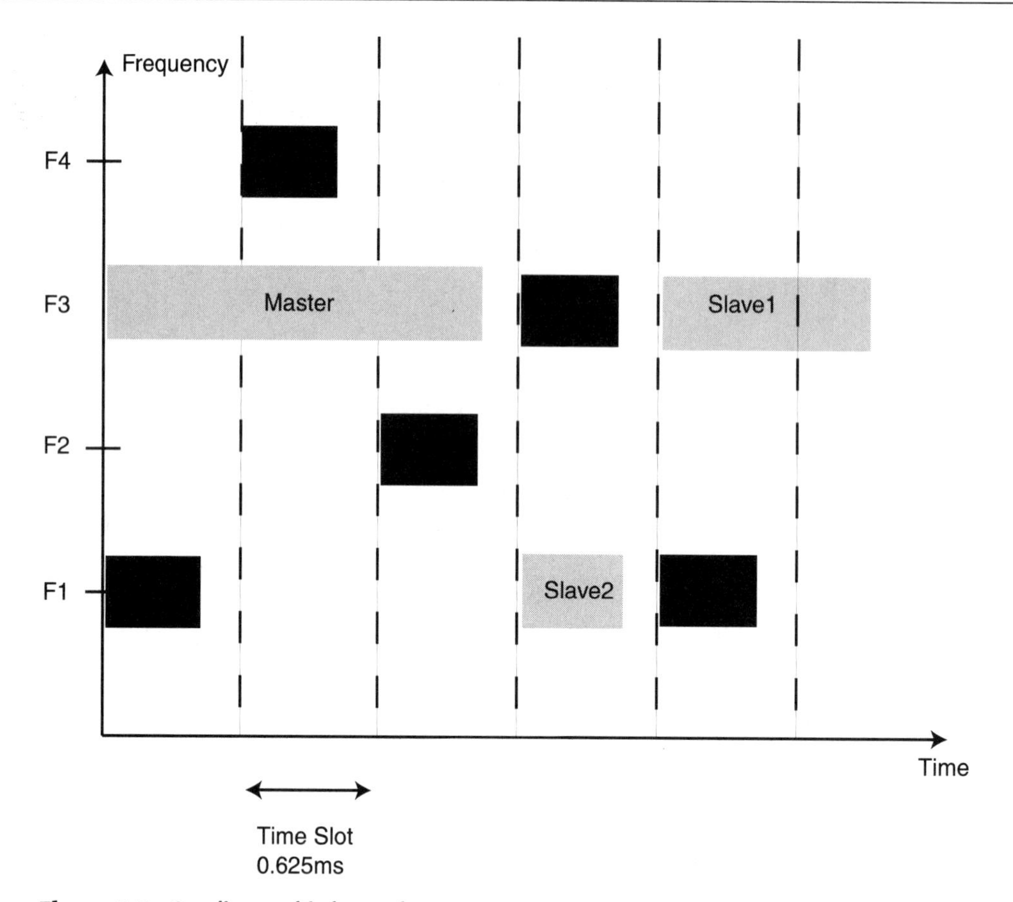

Figure 5.7 Sending multi-slot packets.

that the frequency is unchanged and that the other channel (black packets) keeps the same hopping scheme.

By using multi-slot, bit rates of up to 721Kbps can be achieved, although this rate is without any protective coding. Table 5.1 shows the available bit rates.

DH here means Data High (in other words, with little protective coding and so-called Forward Error Correction, or FEC), while DM stands for Data Medium and uses more FEC. So a link that has low error protection could run applications with 433.9Kbps in both directions or with 721Kbps in one direction and 57.6Kbps in the other. Examples of combined channel configurations include a 64Kbps packet (symmetrical) plus 64Kbps voice (symmetrical, of course) and 3 × 64Kbps voice.

When you are using packet data channels, it is always important to have a method for the retransmission of lost packets. Bluetooth uses an *Acknowledge*

Table 5.1 Available Bit Rates

TYPE	SYMMETRIC (KBPS)	ASYMMETRIC (DL) (KBPS)	ASYMMETRIC (UL) (KBPS)
DM1	108.8	108.8	108.8
DH1	172.8	172.8	172.8
DM3	258.1	387.2	54.4
DH3	390.4	585.6	86.4
DM5	286.7	477.8	36.3
DH5	433.9	721	57.6

Request (ARQ) scheme, which resends packets when it does not receive an acknowledgement that a packet has been safely retrieved. This scheme is similar to what other wireless packet data technologies (such as GPRS) use. By using ARQ, packets that are lost over the air link will not degrade the end-to-end performance significantly. This feature is especially important in dense traffic areas or in places where strong interference is present. A packet that is lost during one slot will soon be retransmitted over another slot. With 1,600 slots per second, a lost packet does not have to be delayed significantly before retransmission. As we will see in Chapter 8, "Adapting for Wireless Challenges," this feature is important when using TCP (which is very sensitive to delays and lost packets).

The Protocol Stack

Because Bluetooth is aimed at serving a large variety of devices and applications, the protocol stack needs to be very flexible. The speakers of your high-fidelity system will likely not have the same TCP/IP stack as your laptop, and it is important to be able to have the best possible protocols for each application. We show the Bluetooth stack in Figure 5.8.

Base-Band

The base-band and link control layer supports the physical *Radio Frequency* (RF) link between Bluetooth devices. This support means functionality of the air interface, such as frequency hopping and synchronization. At this point is where the different link types, such as SCO and ACL that we described previously, are handled. When you are using multiple links (which can be of the same or different link types), they are all multiplexed into the same physical link by this layer.

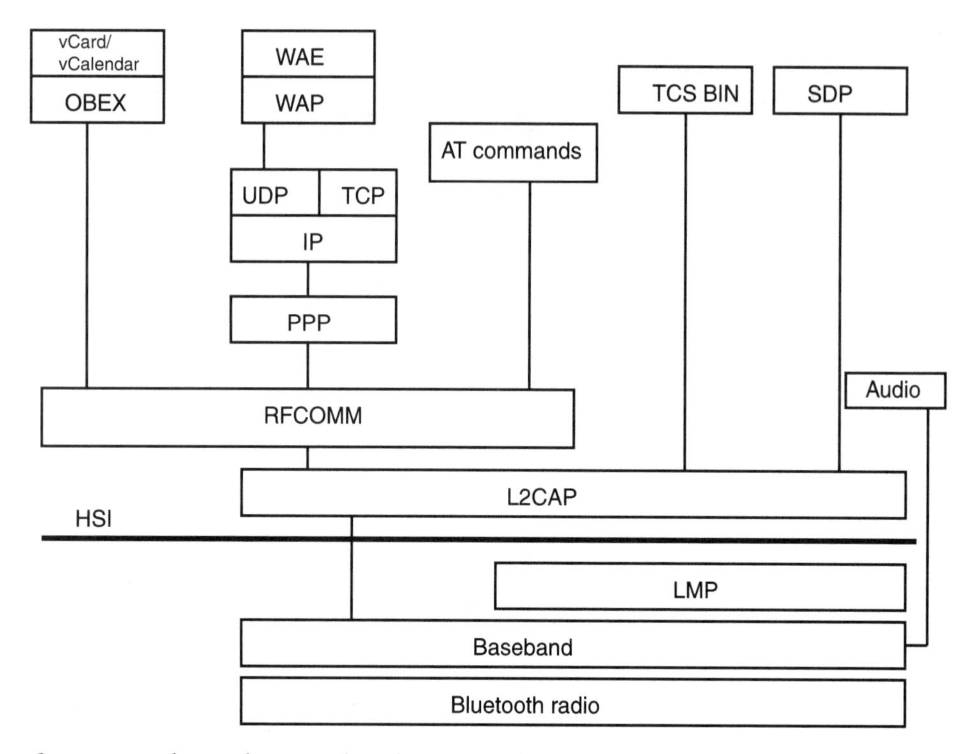

Figure 5.8 Bluetooth protocol stack (source: Bluetooth specification 1.0).

Link Manager Protocol (LMP)

The LMP is in charge of setting up links between Bluetooth units, enabling negotiations of packet sizes, and controlling the negotiations during transmission. In addition, it handles the keys that are involved with security negotiations, power modes, and the state of a unit within a piconet. LMP-layer messages have very high priority and are never delayed by traffic.

Audio

When you are using audio over Bluetooth, there is no need to go through the upper layers; consequently, these profiles use the base band directly.

Host Controller Interface (HCI)

As we mentioned previously, there is a need for various protocol configurations in Bluetooth, but the interface towards the physical layers can be the same. The HCI is exactly that interface, and it provides a uniform method for accessing the physical hardware capabilities. By using this interface, you can access the lower layers and also poll the status of the hardware.

Logical Link Control and Adaptation Layer (L2CAP)

L2CAP is the main interface that connects HCI with the various upper layers. L2CAP provides both connection-less and connection-oriented services and enables easy access to lower-layer functionalities. This functionality includes the multiplexing of protocols, segmentation and reassembly, QoS management, and abstraction of the groups concept in a piconet. In Bluetooth, specification 1.0 does not support SCO links through L2CAP, because those are expected to access the base band directly.

Service Discovery Protocol (SDP)

Because spontaneous networking is a key feature of Bluetooth, there is clearly a need for a protocol that can enable devices to discover each other's services. This feature is implemented by SDP, which defines how a Bluetooth client's application should discover available Bluetooth services. This protocol enables your laptop to find out what that giant monster in the printer room really can do for you (provided that you both are Bluetooth-enabled). This feature is essential for finding and connecting to other Bluetooth devices.

RFCOMM

Most software applications running over Bluetooth will run over a serial port emulation protocol called RFCOMM. RFCOMM enables the use of devices' serial ports and emulates RS-232 control and data signals over the Bluetooth base band. In other words, it acts like a transport mechanism for whatever protocols are used on top of it (perhaps OBEX, TCP/IP, or WAP). RFCOMM corresponds to the IrCOMM protocol in the IrDA protocol stack. RFCOMM runs over L2CAP, and Bluetooth devices will commonly use it. RFCOMM is the fundamental cable replacement protocol that enables a phone to talk to a laptop or to a PDA (and so on).

TCS and AT Commands

The Telephony Control Protocol (TCS) handles the call control signaling when setting up speech and data calls between units. This setup also includes the signaling when releasing those connections. AT commands are used when the software wants to poll or control the hardware, as we described in Chapter 3,

"GPRS-Wireless Packet Data." The Bluetooth AT commands are based on ITU-R Recommendation V.250 and GSM spec 07.07, but the basic usage is the same as for GPRS mobiles.

PPP, IP, TCP, and WAP

Above RFCOMM, it is easy to build a TCP/IP stack or a *Wireless Access Protocol* (WAP) stack on top of a *Point-to-Point Protocol* (PPP). Although RFCOMM emulates a serial connection, this stack is used for ad-hoc Bluetooth *Local Access Networks* (LANs). This method is also the way in which an Internet bridge (described next) is created, enabling a laptop to access the Internet via a Bluetooth-enabled cellular telephone. RFCOMM creates a base on which any of the usual configurations with these protocols can be used.

Bluetooth Networking and Profiles

In order to ensure that Bluetooth devices from different manufacturers will work together, it is important to specify how they should interoperate. This is done through a number of usage models and profiles. A profile describes basic properties, such as protocols, messages, and procedures, that implement a feature. All of these properties are then shared among a number of Bluetooth units of a usage model. There are four general profiles: *Generic Access Profile* (GAP), *Service Discovery Application Profile* (SDAP), *Serial Port Profile* (SPP), and *Generic Object Exchange Profile* (GOEP). The generic profiles specify those features that are common to several usage models, such as how to discover other Bluetooth units. When describing the individual usage models (such as the ultimate headset), a number of specific profiles are used, in addition to the general ones. In this way, a user model is supported by a number of general and specific profiles. As an example, the Ultimate Headset user model is supported by the general Serial Port profile and by the specific Headset profile. In other words, a product that uses this usage model can communicate with a phone that supports the Headset profile, for instance.

Before digging into the profiles and user models, we need to know a bit more about Bluetooth networking.

The Piconet

Bluetooth devices can interact in a number of different ways, as we will see later in the section about usage models. Any two devices that are within range of each other can set up a so-called ad-hoc connection, which means that they form a piconet. One unit (usually the one that initiates the piconet) becomes the master, and the rest act as slaves. The master is the one that chooses the frequency-hopping sequence and that is in control of the transmissions. The master controls all traffic in the sense that it allocates capacity for SCO links and decides who can transmit on ACL links. In order for a slave to send in the so-called slave-to-master slot, it has to be addressed in the preceding master-to-slave slot. Consequently, slaves can only receive information in a specific order and send upon being addressed. This technique helps avoid packet collisions between multiple slaves that are sending at the same time.

There is, however, no difference in hardware and software between masters and slaves, and a device that is a master in one piconet can be a slave in another. The actual functionality is, however, decided by the profile in question and some, like the headset, are not likely to ever be used as masters. Roles can change in a piconet, but there is never more than one master. Up to eight units can share the same piconet, which effectively means that they can share the same 1MHz channel and corresponding hopping frequency. There can actually be more than eight members of a piconet but rather 200 as long as only eight are active at the same time. The others are then inactive and can awake when it is time to communicate. The fact that all members of one piconet are sharing the same 1MHz channels causes the overall gross capacity for all of them to be limited to the usual 1Mbps. So, if there are eight active users, bit rates will be significantly lower. A remedy to this problem is to use several piconets in a configuration that is called a scatternet.

The Scatternet

In order to maximize the capacity of ad-hoc networks, several piconets can be connected in a scatternet. One user is a member of several piconets, as seen in Figure 5.9 (where the phone is in both piconets).

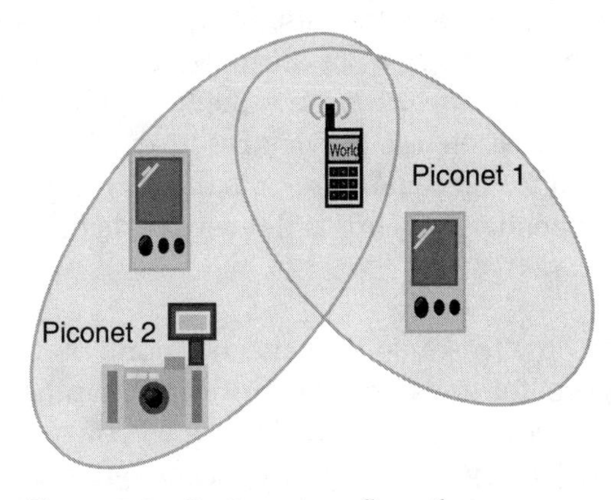

Figure 5.9 Scatternet configuration.

Each piconet uses a different 1Mhz channel and 1Mbps gross total bit rate that is shared among the users in the piconet, while users from different piconets do not have to share capacity at all. So, in order to get the highest possible throughput for all users, the scatternet should consist of as many piconets as possible. The drawback is that in total, there are only 79 available frequencies, and the more piconets that are allocated, the more frequencies are used. The laws of probability imply that the number of collisions resulting in retransmission is so low that up to eight piconets are possible in one scatternet.

Defining General Profiles

The profile concept originated in order to organize different Bluetooth devices and applications into categories that enable interoperability between units from different vendors. As we mentioned previously, there are four general profiles that specify properties that are common to several usage models.

Generic Access Profile (GAP)

The GAP is the mother of all Bluetooth profiles and defines how two units discover each other and establish connections. The profile defines a number of operations that are generic and that specific profiles can use, much in the same way that objects in software can inherit objects from ancestors. GAP ensures that Bluetooth devices from different vendors can discover each other and start communicating. A Bluetooth unit that does not conform to any other profile must still conform to GAP. Therefore, the main purpose of GAP is to define how the lower layers of the protocol stack (LC and LMP) are used. As the ancestor of all other profiles, it also describes how to handle several profiles simultaneously and coordinate them. GAP also includes the different operation modes, such as discoverable and standby states.

Service Discovery Application Profile (SDAP)

While GAP enables the discovery of Bluetooth units, SDAP discovers services. The search can either be for specific services or attributes or generic service browsing, looking for any service that the other party might support. This general profile also includes an application, the Service Discovery User Application. This application is required for all Bluetooth units in order to enable the locating of services. SDAP is dependent on GAP and reuses parts of it.

Serial Port Profile (SPP)

The serial port profile defines how to set up virtual serial ports on two Bluetooth devices and use those ports for transferring information (in other words, setting up a connection-oriented channel). The two devices will then communicate through an emulated RS-232 link, which means that the usage of this profile is similar to what developers already know. Many legacy applications can run over Bluetooth as a result, because they think that it is a regular serial cable. To perform this task, a Bluetooth helper application is used to set up the Bluetooth stack, but no modifications to the actual application should be necessary.

Generic Object Exchange Profile (GOEP)

The GOEP builds on the SPP and enables the exchange of objects between Bluetooth units. These objects can be files (File Transfer Profile), Personal Information Management (PIM) data (Synchronization Profile), business cards, and so on. Laptops, mobile phones, and *personal digital assistants* (PDAs) will commonly support this profile. GOEP was designed for ease of use across different platforms (for instance, enabling the exchange of business cards between phones and PDAs).

Specific Profiles for Usage Models

The specific profiles are the descriptions of features and processes that need to match in order to guarantee interoperability between Bluetooth devices (see Figure 5.10). That is why the developers describe these profiles in great detail in the Bluetooth specification and have an important role for the flexibility of Bluetooth. All of the specific profiles of Bluetooth 1.0 appear as follows, but the deeper details are left out. The aim is to improve the understanding of Bluetooth technology and its features, and those who need more details should download the specification (from www.bluetooth.com).

Cordless Telephony Profile

Because Bluetooth has highly developed capabilities for voice, it is natural to use it as a small office or home cordless phone. The Cordless Telephony Profile defines how to make calls from a phone via a home base station to the public phone network, just like existing cordless systems such as Digital Enhanced Cordless Telecommunication (DECT).

Intercom Profile

In addition to the possibility to use Bluetooth for cordless phones, the phones can be made to act like walkie-talkies with two phones talking to each other without going through any (home) base station. The Intercom Profile defines this phone-to-phone connection feature.

Headset Profile

The cables that are used for wireless headsets have a tendency to get tangled in ways that seem incomprehensible even for the best physicist, and enabling wireless headsets was one of Bluetooth's prime objectives. The Headset Profile

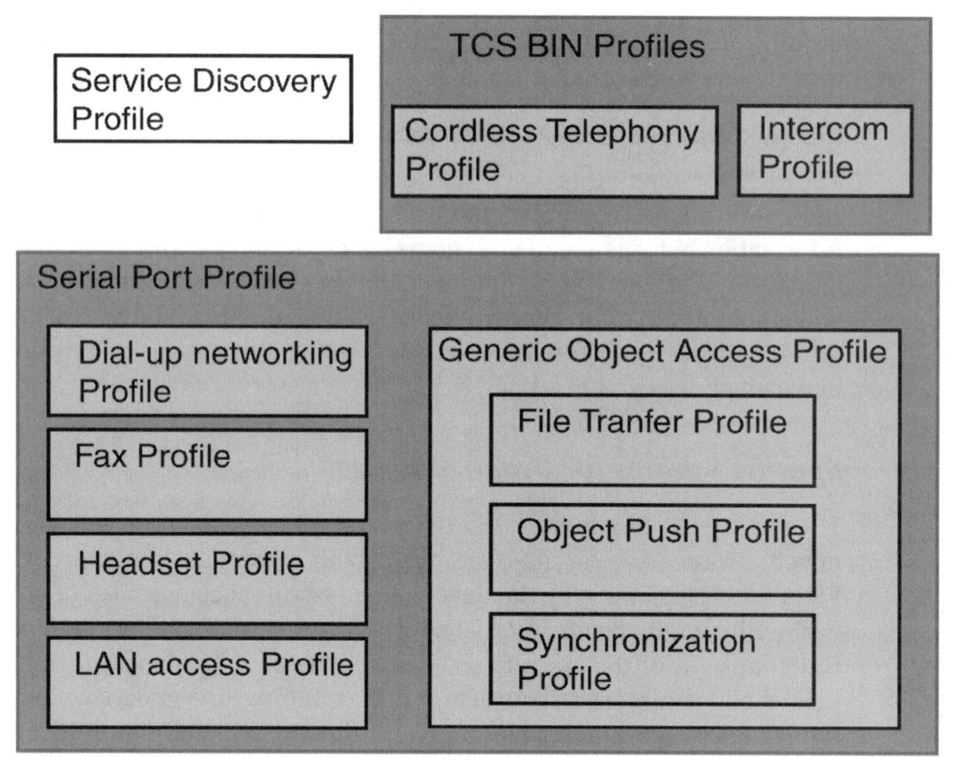

Figure 5.10 Profiles and their relations
Source: Bluetooth specification 1.0.

describes how you can use such a headset with mobile phones as well as with personal computers and other devices. The headset profile is based upon the Serial Port Profile.

Dial-Up Networking Profile

In order to access online content from a laptop via a Bluetooth-enabled phone, you need support for a regular dial-up connection. The Dial-Up Networking Profile provides this functionality and is based on the Serial Port Profile. Just like a regular modem, the laptop (or whoever uses the service) uses AT commands to control the setup and maintenance of the link.

Fax Profile

In the same way that laptops and PDAs can use the Dial-Up Networking Profile to access modem functionality, the Fax Profile offers fax capabilities to other Bluetooth-enabled devices. This profile is based on the SPP.

LAN Access Profile

We expect that LAN access via Bluetooth will become very popular because it will enable devices to connect wirelessly to a nearby LAN access point. This connection is facilitated by the LAN Access Profile, which uses PPP over RFCOMM and enables access to a wired LAN through an access point and initiates the LAN setup between individual Bluetooth devices. With WAP over Bluetooth, you can use a WAP phone to control a PC, and there will be many interesting applications of this feature.

File Transfer Profile

The File Transfer Profile not only supports the transmission of files and documents but also controls another device remotely in order to access files and create and delete folders. The File Transfer Profile is dependent on the GOEP.

Object Push Profile

The broad success of SMS in Europe and business card exchange by using infrared connections has proven the importance of being able to push content to other users. In Bluetooth, this feature is handled by the Object Push Profile, which describes how to push objects between devices. This profile is most commonly used among laptop PCs, mobile phones, and PDAs. Examples include sending a business card to someone or sending them information about a meeting.

Synchronization Profile

The rapidly increasing use of PDAs and the possibility of synchronizing them with PCs creates the urge to synchronize wirelessly. The Synchronization Profile defines how this process can be done in Bluetooth and creates the potential to perform this task across many different devices. The possibility of automatic synchronization exists as soon as two devices come close to each other, making it possible for users to have all devices synchronized without having to remember to run the synchronization program.

Examples of Usage Models

Because the specific profiles define the least common denominator for different Bluetooth units, most devices will support several profiles. Here are some examples of such usage models, but for a complete description, please refer to the Bluetooth specification.

Internet Bridge

Profiles: Dial-up networking profile, fax profile, LAN access profile, SPP

The divided concept, described more in detail in Chapter 10, "Mobile Internet Devices," describes how many users will use the phone as a modem and then use a laptop or PDA for the applications. Bluetooth is an excellent facilitator of this usage model, and the Bluetooth-enabled phone then acts as an Internet bridge. The phone should then have both Bluetooth and GPRS/3G functionality (using the Bluetooth connectivity between the phone and PDA/laptop). This process requires a two-piece protocol stack—one for the actual data and one for AT commands that can control the mobile phone (like a modem, in this case).

The Ultimate Headset

Profiles: Headset profile, SPP

The Ultimate Headset describes how a Bluetooth headset can be used to not only send and receive voice packets, but also to command the phone to (for instance) answer and terminate calls. As with the Internet bridge, this functionality requires a two-piece protocol stack: one for the actual data and one for AT commands that can control the mobile phone. Note that unlike regular mobile phone headsets, a Bluetooth headset can be used with phones from multiple vendors and also with PDAs and laptop computers.

Profile Example in Existing Bluetooth Products

When you specify the capabilities of Bluetooth products, you will likely use the profiles. The official Bluetooth Web site, www.bluetooth.com, lists the latest-released products in a number of categories. Toshiba released its first version of an Internet bridge called a Bluetooth Modem Station. The description is as follows:

Toshiba Bluetooth Wireless Modem Station

PROFILES:

- Generic access profile
- Serial port profile
- Dial-up networking profile
- Fax profile

Compare this description to the previous section, and you will find that the description is similar to the Internet bridge usage model. Another example is a PC card from Digianswer:

Digianswer Bluetooth PC Card

PROFILES:

- Generic access profile
- Service discovery application profile

- Serial port profile
- Dial-up networking profile
- Fax profile
- LAN access profile
- Object push profile
- File transfer profile

The PC card consequently is a much more versatile device and is expected to interact with many different kinds of Bluetooth units. The Bluetooth home page is a good source for information about the compliance of released Bluetooth units that can be very helpful when designing applications around them.

The Applications and User Interfaces

We often use the word "application" in many different contexts and for different things, as mentioned in Chapter 1, "Basic Concepts." Throughout this book, we talk about software applications and services—but for Bluetooth, there will be many combined hardware/software products that are of great importance. Because there were initially no Bluetooth-enabled devices for developers to play with, the first wave of applications will be of the hardware/software kind (and often standalone applications). As more and more Bluetooth devices appear (and some even with open platforms, such as Communicators), the software can enter the scene on a wide scale and provide a wide range of interesting applications. Bluetooth is one of those technologies that opens up such interesting possibilities that even its creators have no idea what the innovators will come up with as uses for it.

The first Bluetooth implementations are likely to focus on point-to-point communication, such as a wireless headset for mobile phones, PC cards for laptops that can access the Internet through Bluetooth-enabled mobile phones, and so on.

An important question is how Bluetooth will look and feel for the user. We do not mean the radio waves; rather, we are talking about the *Man-Machine Interface* (MMI) for devices. This issue is really two-fold in the sense that you have one interaction when new devices and services are discovered and one when the service is being used. We describe a few user models and the MMI in the following paragraphs.

Headset MMI Example

When a headset is to be used with a mobile phone, the two units need to discover each other and become paired. The following procedure outlines a common way of performing this task (used by the Ericsson Bluetooth headset).

1. The button on the headset is pressed for 10 seconds so that the light starts blinking red and green. This means that it is ready to get paired.
2. On the phone, the Bluetooth menu alternative is selected, and you can then select to pair the phone with another device.
3. You are then asked to prepare the other unit (here, the headset) and press a button to start the discovery process.
4. As the phone is finished with the discovery process, it displays the available units. The headset can then be selected.
5. The phone indicates that the devices are now paired, and you can use the headset.

In order for this process to work, both units have to have the same four-digit *personal identification number* (PIN) code (which usually is 0000 as a default for both). When you pair the two, the headset can pick up incoming calls with a click on the attached button, and you can adjust the volume. The voice command feature also enables voice-activated calling. This example is typical and shows how two different protocol stacks are used dynamically. The AT commands enable the headset to control the phone, and the audio stack enables the actual sound transfer.

Bluetooth-Specific Development Considerations

Everyone seems to agree that there are endless possibilities with Bluetooth and that it will be a major enabler of the mobile Internet. As more and more devices get the chips built in from the start, it will be just as common as digital clocks (which you see today in everything from laptops and phones to TVs and billboards). Maybe Bluetooth will be even more pervasive? For applications developers, the question is where to approach this giant opportunity and what to think about. At this writing, the main issue is what platform to use for software applications. Hardware developers are less concerned about this issue, though, because they can more or less create their own platforms (but that is out of the scope of this book). Software developers, on the other hand, are very much at the mercy of device manufacturers and platform developers. There are still a number of key considerations that we can see even at this very early stage.

Device Agnosticism

Say for instance that you have developed a Java-based chat application that uses the spontaneous networking features of Bluetooth. Now, laptops, PCs, and high-end PDAs (assuming that personal Java is used) could potentially use this application. The number of devices and screens on which this application could run is very large and will grow quickly. You will have a clear advantage if most of the application is made independent of form factors and input mechanisms. This statement goes for many Bluetooth applications where the number of potential platforms can be huge.

Ease of Use

You might think it obvious that any application should be user-friendly and easy to start, but these features are especially important with Bluetooth applications. This technology introduces many new usage scenarios, and changing people's behaviors and mindsets is not easy. To explain this rather vague advice, let's compare a pre-Bluetooth scenario with the unwired alternative. In the late 1990s, many people used a portable, hands-free device with their mobile phones—basically a wire from the phone to the earphone with a microphone on the end. They could use the phone in either hands-free mode or in normal mode, and it was obvious which mode they used (when the cord was attached to the phone, hands-free mode activated). If you wanted to talk directly into the phone, you would simply remove the phone. This usage pattern was the same for other accessories, too, such as chatboards, mp3 -players, and so on—snap it on and use it or take it away when you do not need it. With Bluetooth, this usage becomes a bit less intuitive because there is no longer a physical connection between the devices that talk to each other. A wireless headset does not look much different when it is paired with the phone than with the laptop (or not paired at all). This scenario will be a massive change in user behavior and the way in which people view connectivity. People should take this change seriously, and this change will not happen overnight. As an application developer, it is important to realize this fact and to be over-communicative of how devices are connected and interact. We do not want to come to a situation where users feel powerless and out of control.

Security and Comfort

Having all sorts of devices talking to each other will appear like magic to many people, and this situation can lead to discomfort among users. If my laptop easily discovered my printer without my interaction, how do I know that no one discovered me and started eavesdropping on our communication? No matter how much we discuss and analyze the security aspects, it comes down to two things: making things secure enough and making users feel comfortable. Because most of the security is built into lower layers, this situation might be hard for applications developers to control. Often, it is enough to keep the user informed (for instance, about the trusted device that the application has found). Also, one could add application-layer security measures, which usually are visible to the user (certificates and so on). This situation usually brings you to the same point of tradeoff between security and convenience: ease of use.

Summary

Bluetooth is a short-range radio technology that enables devices of many different shapes and functionalities and from different vendors to communicate. The technology operates in the 2.4GHz band and uses frequency hopping in order to ensure robustness against interference in this open frequency band. Profiles help ensure the interoperability between devices. Profiles indicate what functionality a device supports, and devices that feature the same profiles can communicate. Bluetooth is designed to be small and power-efficient and is expected to be included in a huge variety of devices (such as mobile phones, PDAs, computers, and so on).

PART TWO

Optimizing the Transmission

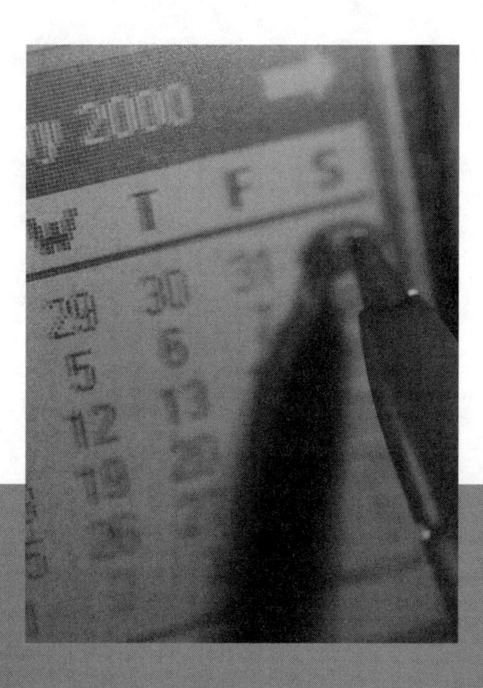

CHAPTER 6

Unwiring the Internet

Many people initially viewed the mobile Internet as the Web without wires, meaning the same content and applications but without any fixed, wired connection to the user's device. While this perception led to the creation of some flashy commercials and scenarios that people could relate to with their past Internet experiences in mind, it also created false expectations among users and wrong approaches among developers. Some companies just took the same applications and content that they presented to fixed Internet users and brought it to mobile Internet users. This approach was as bound for success as scanning the company archives to be presented on the Web in the early 1990s.

When the Internet as we know it today was built, its building blocks were constructed to work well with the network characteristics that were most prominent at the time. We could assume that a link had pretty much the same properties all of the time unless it got too loaded (congested). In case of congestion, all hosts on the network should act responsibly and back off to get the system out of the congestion. In this way, protocols such as the *Transmission Control Protocol* (TCP) and the *Hypertext Transfer Protocol* (HTTP) were designed with the Internet of past decades in mind. In TCP, a lost packet is treated as sign of congestion (that too many users are trying to use the network at the same time). Therefore, TCP backs off and reduces its transmission speed when it notices lost packets. These are some of the issues that we will address in this chapter, and we will explore some concrete ways to avoid getting into trouble when taking existing applications into the mobile Internet world. This process is not only about adapting the content, but also about adapting to the properties of the new networks.

Background and History

In the late 1960s, the United States Department of Defense and the *Advanced Research Projects Agency* (ARPA) established partnerships with U.S. universities and company research divisions in order to create a community for information sharing over computer networks. This goal required open, standardized protocols and a distributed multi-vendor architecture.

They created the first embryo of the Internet, called ARPANET, which was a packet-switched network with not-so-impressive bit rates of 56Kbps. This first network, which launched in 1969, consisted of four nodes and did not look much like the Internet that later would appeal so highly to the public. During the 1970s, this community of bright people worked on the architecture and protocol issues surrounding this network. In 1974, Vinton G. Cerf and Robert E. Kahn wrote a paper that outlined a design of a new protocol suite for the Internet. The proposal evolved into what we commonly call the TCP/IP Internet protocol suite, popularly just called TCP/IP after two of its components, TCP and the *Internet Protocol* (IP). After a few initial versions, work ended up with the commonly used version 4, which they finalized in 1979. IP version 4 is still, in the early twenty-first century, the protocol used on almost every computer on the Internet.

As more and more universities and other nodes became connected, capacity was running out, and the network was constructed piece by piece. In the late 1980s, T1 links of 1.544Mbps emerged, and T3s soon followed in early 1990. The universities continued to lead the Internet evolution as the use of e-mail became widespread in the early 1990s and was not only limited to researchers and computer students.

In the early 1990s, the Internet received a new face with the rapid spread of Web browsers that supported *Hypertext Markup Language* (HTML) over HTTP. Mosaic and Netscape were the early pioneers, and Internet Explorer later followed. Now, the common computer user could get a modem fairly cheaply and access the massive amount of information that was available. Again, universities led the way—and for every generation of students that graduated, the power of the Internet became more pervasive. With the takeup among consumers, people started to think about how to make money on this emerging opportunity.

In the second half of the 1990s, commercial interests became a bigger and bigger part of the Internet. People started to book trips and buy books and other goods online, and the Internet became an integral part of many people's lives. From having been something that universities used for communication and research less than 10 years ago, the Internet was now a natural part of work as well as spare time.

The Internet Protocols in Wireless

The advent of the mobile Internet in the late 1990s created content that people could previously only access at certain fixed locations but now was potentially available everywhere. While the thought of reusing as much as possible of the existing Internet infrastructure and protocols was beneficial in many ways, it also created some problems. The protocols and content were created with user models in mind that were not always applicable to mobile users. Although many of us will use our laptop to access the same content wirelessly as from our desktop PC, the majority of the mobile Internet use is from much smaller handsets and with slower links. Therefore, the mobile Internet application developer needs to be aware of how the Internet protocols behave in a wireless environment.

The OSI Model for the Internet

After using TCP/IP for a while, people realized the importance of a flexible protocol stack. In the mid-1970s, the *Open Systems Interconnect* (OSI) model emerged. This model divides communication into layers. Each layer fulfills certain tasks and makes different combinations for different applications. While the OSI model might be too basic for advanced developers, we will quickly refresh your memory. Figure 6.1 shows the generic OSI model and how we can interpret it for the Internet.

The different layers of the Internet model are as follows:

Network interface layer. This layer is where the actual bits are transported and the hardware addresses (such as Ethernet) for the physical host computers are specified. The network interface layer formats packets and sends them via the underlying network. For the mobile Internet, this layer includes the air interface.

Internet layer. This layer is equivalent to the network layer in the OSI model and primarily includes IP. IP addresses make it possible to locate the destination host and to send packets to it without having to be on the same subnet. A *Domain Name Server* (DNS) makes it possible to translate an easy-to-remember address, such as www.ibm.com, to the IP address that specifies where the IP packets should be sent.

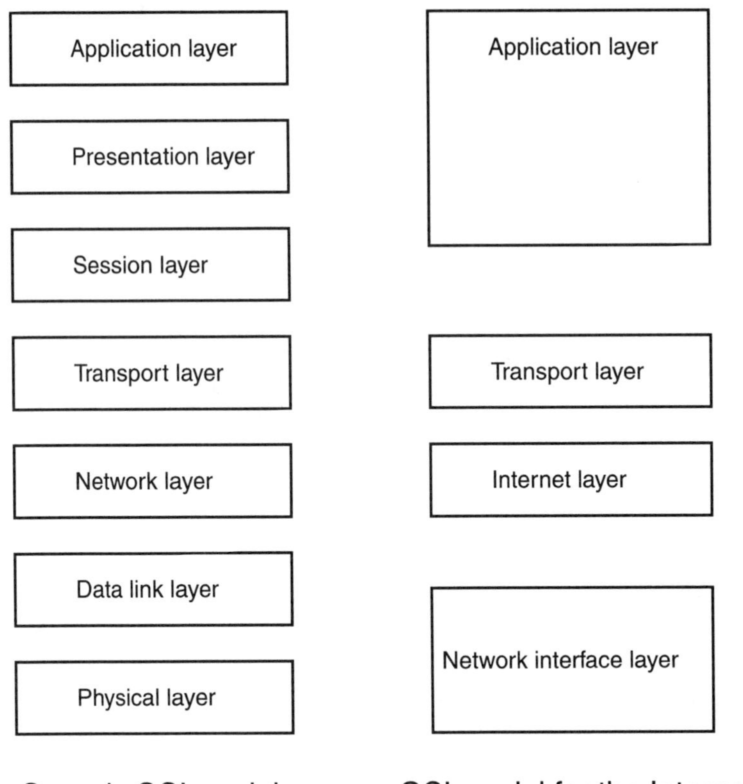

Figure 6.1 The generic OSI model compared to the Internet model.

Transport layer. Now that a packet has reached the host computer, it needs to know which application we want. A port number of 80 specifies that we want to talk to the Web server (HTTP), and 21 specifies a *File Transfer Protocol* (FTP) server, and so on. Another important feature of the transport protocols is the capability to deliver the packets reliably, in the right order, and at the appropriate speed. TCP performs all of these tasks while User Datagram Protocol (UDP) just delivers the packets as they come (without caring about how many get through). This feature is good for real-time applications, however.

Application layer. The user usually is not faced with any parts of the protocol stack except for the application layer protocols. Here, we find FTP, HTTP, and other protocols that format the content and deliver it. The application layer corresponds to all three layers at the top of the OSI model: the session layer, the presentation layer, and the application layer.

Internet Protocol (IP)

IP transports packets to the desired destination host on the network. IP is a connectionless protocol and is not aware of any sessions. Every packet is routed independently, and different parts of the same transmission might take a different route. Along the way a packet might be lost, corrupted, duplicated, or delivered out of sequence (in other words, the first packet that is sent might not be the first one that is received). If the underlying network is not capable of transmitting packets as large as those that higher layers try to get IP to send, IP will fragment the packets in order to fit the network. For example, if someone tries to send a 2,300-byte packet over an Ethernet network that only can handle packet sizes of 1,500 bytes or fewer, the 2,300 bytes will be segmented into two IP packets. Incidentally, IP sends packets by using the best-effort principle, and whatever gets lost or received out of sequence is the responsibility of the higher-layer protocols.

IP is definitely the future, and we will see mobile networks comprising an ever-increasing degree of pure IP. For a protocol that was specified in the late 1970s, IP has proved to be a versatile and scalable part of the TCP/IP protocol suite. The main challenge with IP today is that the currently used version, IP version 4 (IPv4), has too limited an amount of available addresses. Just with the growth of the fixed Internet, the addresses are predicted to run out rather soon. The lack of IP addresses is already today a limiting factor for the growth of some applications. With the advent of the mobile Internet (and its predicted 600 million users in 2004), it will be tough to find any free addresses. Also, today the distribution of IP addresses is not really fair (with one university getting more IP addresses than the entire Republic of China).

The solution, IP version 6 (IPv6), was developed in order to ensure that IP would not become a limiting factor for the spread of the Internet. With IPv6, 128 bits are used for the IP address instead of 32 bits (see Figure 6.2). Consequently, there will be a theoretical limit of 340×10^{1038} Internet hosts (or plenty of billions of addresses per person on Earth). In addition to solving the problem of the lack of IP addresses, IPv6 also provides enhanced security features and more control over the routes that packets take.

IPv6 has been in the works for a long time now, but adaptation is slow in the industry. We expect, however, that the mobile Internet will be one of the main drivers for a wider acceptance of IPv6.

While IPv4 still exists, static IP addresses are likely to come at a premium. Most wireless networks will enable the user to connect either to an *Internet Service Provider* (ISP) that the wireless operator provides or to an external ISP (through a RADIUS/DIAMETER server). Anyway, we do not recommend

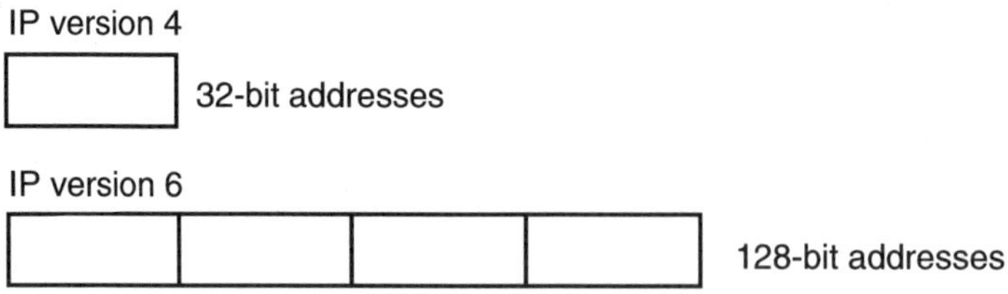

IP version 4

32-bit addresses

IP version 6

128-bit addresses

Figure 6.2 IP address sizes for version 4 and version 6.

designing an application that requires static or public IP addresses, because not many users will be able to obtain one of those addresses.

Transmission Control Protocol (TCP)

As a transport-layer protocol, TCP also ensures that data for which IP finds a destination host will be propagated to the right application. FTP and HTTP applications might be running on that same host, and the TCP port number indicates which is the target application. TCP also ensures that packets (TCP really sends and receives segments, but here we use the word *packet* at all times for simplicity) of data are delivered reliably and in order. As we saw previously, these requirements are crucial because IP does not guarantee anything; rather, it merely provides a means to get the packets routed correctly. Packets that have traveled different routes in order to get to same destination need to be assembled in the right order, and TCP performs this task. Delivering the packets reliably means that lost packets are detected and retransmitted. Finally, TCP provides flow control functionality, which makes sure that the sender and receiver agree on a suitable speed of data delivery. Covering the loss of packets and the maintenance of flow control are the two features that affect wireless performance the most. In addition, applications developers who want to get the most from this versatile protocol should understand the setup process of a TCP session.

Establishing a TCP Session

TCP is a connection-oriented protocol, which means that a session has to be initiated before data can be exchanged and that each TCP session can only exist between two hosts. The typical example is a user who wants to download a Web page with his or her Web browser. Clicking a hyperlink starts a TCP session between the user's client computer and the Web server that hosts the page. As the user goes to another site, a new TCP session is created between that server and the client. Each TCP session is initiated by a three-way handshake, where the initiating party (let's call him or her the sender) sends a *synchronization* (SYN) packet to the other party (here, we call it the destination). This packet holds information about the session that is to be established, such as the packet sequence number, the proposed maximum packet size, and the proposed rate of transmission. The rate of transmission is identified by a sending window size (in other words, how big a part of the sending buffer can be sent at the same time). The receiving party then responds with similar parameters that consequently describe its parameters and buffer sizes. As a final step, the sending party acknowledges that it received the responding SYN by sending another SYN packet. We show the entire sequence in Figure 6.3.

So, every time you establish a TCP session, you exchange three packets. This is not a serious problem on the Internet, because the latencies (delays between the sender and the receiver) are low (200ms to 300ms most of the time). In a

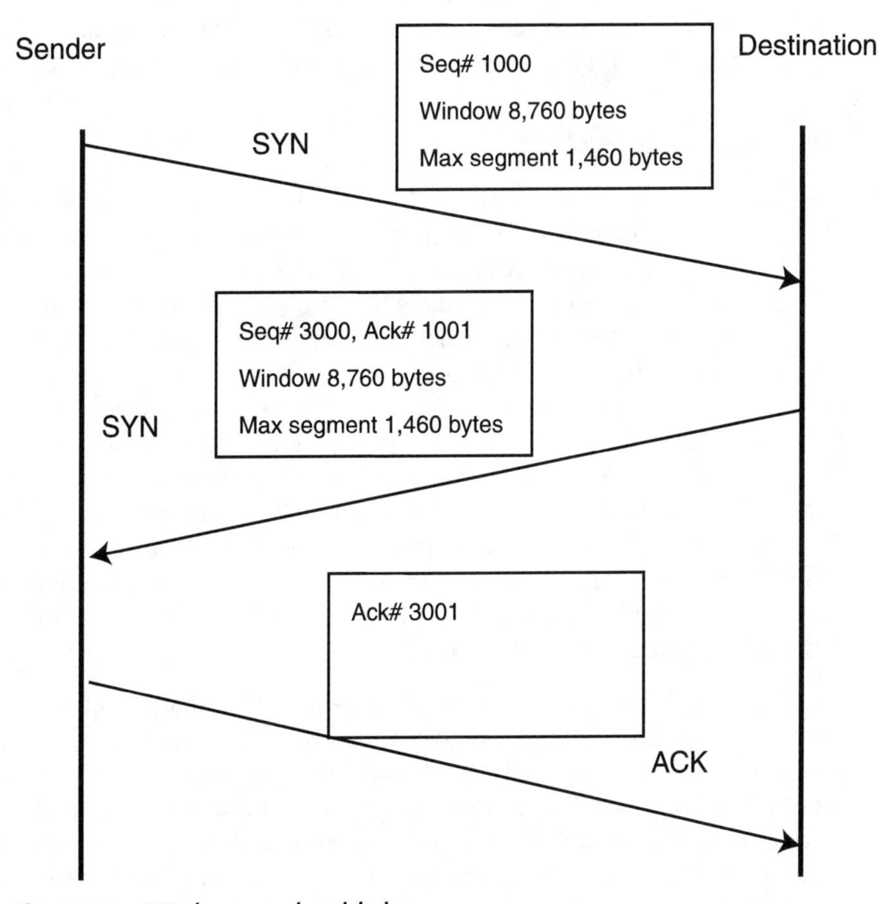

Figure 6.3 TCP three-way handshake.

wireless system, however, latencies will (at times) be in the range of seconds. Think of what the consequences are on a TCP network that has a *round-trip time* (RTT) of two seconds, because each transaction between the sender and the receiver needs packets to go back and forth a number of times. RTT refers to the time that it takes for a packet to travel to the recipient and back. If you have to use TCP, be aware of this factor and always count on a delay that is measured in seconds when establishing a new TCP session.

In addition, there is a parameter called *Initial Retransmission Timeout* (IRTO) that specifies the time that a sender will wait for the first sent packet to return before assuming that the other party will not respond. The *Internet Engineering Task Force* (IETF) recommends this parameter to be set to three seconds, but many servers on the Internet have cut that down to a few hundreds of milliseconds. Why? Well, if you are on a fast connection and are accessing a server, you do not have to wait that long before abandoning the connection if it is unreasonably slow or even down. Obviously, an RTT of two seconds (not uncommon on loaded wireless networks) here would cause considerable problems, because the sender will give up the wait for the initial packets even before they have traveled halfway.

Resending Lost Packets

Packets that travel over any network (not only the wireless ones) will occasionally be lost along the way. This can be due to overloaded routers or just physical disturbances in the transmission. TCP provides a means for overcoming this problem by resending packets that are lost or severely delayed. Timeouts and packet acknowledgements help track the losses of packets. As we saw in the previous example of establishing a connection, TCP responds to a received packet with an acknowledgement packet. This packet indicates the order of the last received packet. When the sender sends a packet, it starts a timer that indicates how long it has taken for the packet to reach the destination and how long it took for the destination to send an acknowledgement back. When the sender's timeout expires, the packet is resent. In Figure 6.4, we show a simplified example of retransmission caused by timeouts. Here, we see that only one packet is sent, while in reality, TCP will send several packets while waiting for acknowledgements.

Sometimes it can be good to detect losses earlier than what the timeout enables by tracking the acknowledgements. By using cumulative acknowledgements, the destination can indicate the last packet that was successfully received. Then, if a packet is not received, the next one in line will arrive before it. To indicate that a packet is missing at the destination end, the last packet in sequence that was received is acknowledged. Figure 6.5 illustrates how losing packet 2 makes the recipient send a duplicated acknowledgement of packet 1, because that was the last-in-order packet that it received.

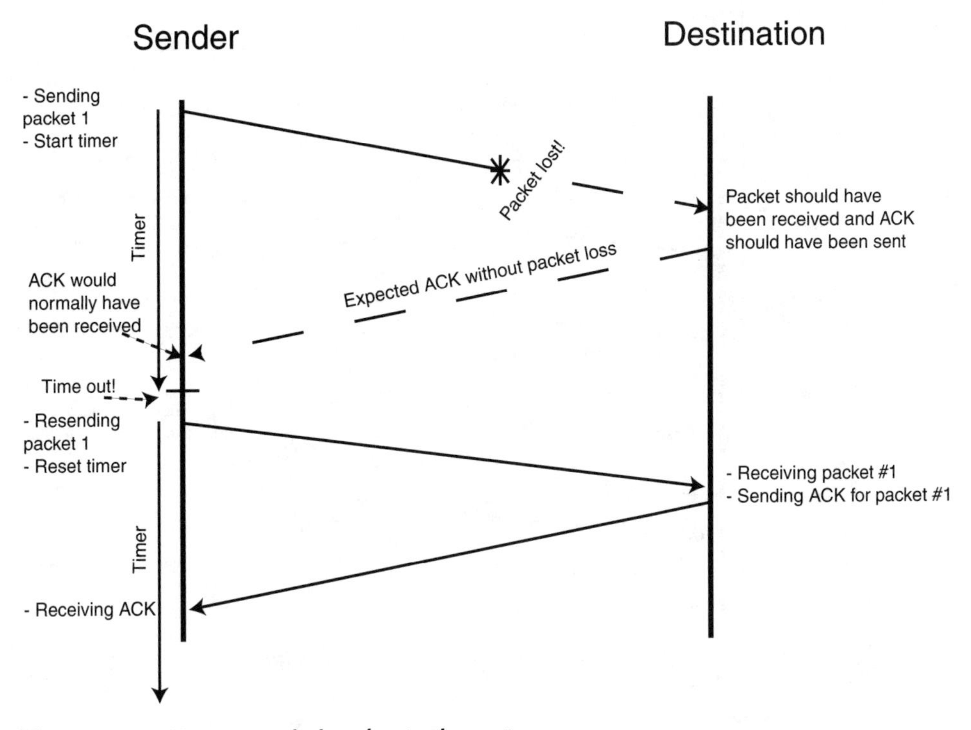

Figure 6.4 TCP retransmission due to timeout.

As the sender receives this indication, he or she knows that other packets have arrived before packet 2—possibly indicating that this packet was lost. Packet 2 is therefore resent, and the sender does not have to wait until the timer expires. By detecting double acknowledgements in this way, TCP can detect lost packets and resend them more quickly.

If we now look at this retransmission scheme in a wireless system, however, there is a potential problem. TCP facilitates reliable connectivity between two hosts, end-to-end. In a typical scenario with a user in New York surfing a Web site that is based in California on his GPRS-enabled laptop, the wireless link might be 1km out of a total of 1000km. Because the wireless link is subjected to disturbances and losses to a much larger magnitude, covering losses with an end-to-end protocol might not be the best idea. Most wireless systems, therefore, have their own link-layer retransmission protocols that can operate over the air link only and quickly resend lost packets. In other words, the TCP session will not notice that packets got lost over the air. As an example, *General Packet Radio Services* (GPRS) uses the RLC protocol over the air interface in order to hide the packet loss from the end-to-end protocols. The retransmission over the air link then removes most of the packet loss, but the price paid is

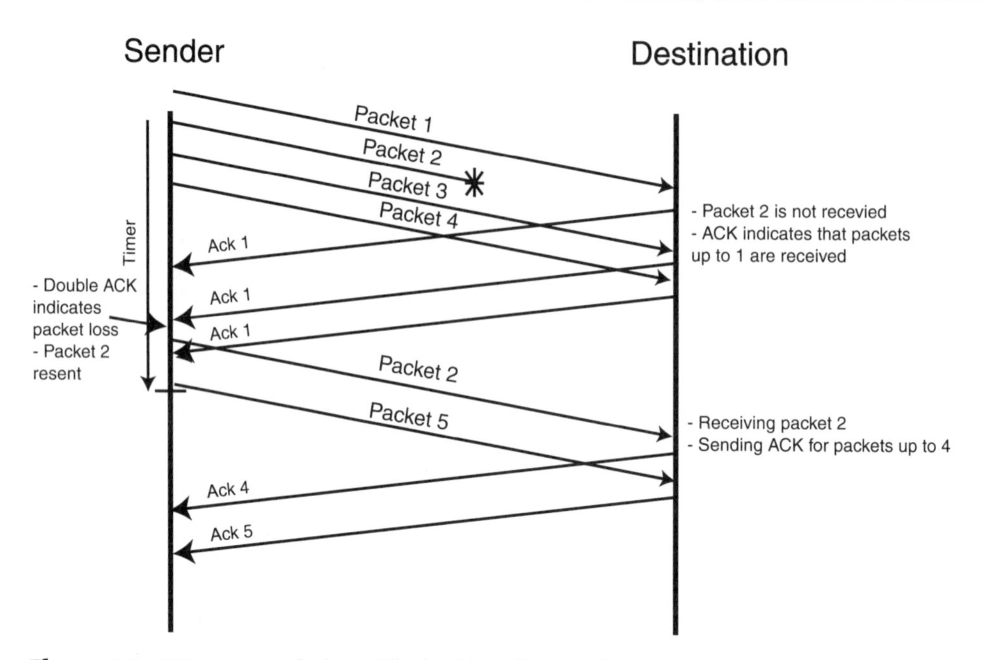

Figure 6.5 TCP retransmission with double acknowledgments.

the delay that the retransmissions cause. In cases where the bit-error rate is high, the latency that the retransmission protocol adds will be significant. In CDMA systems, where the target signal quality can be adjusted in the system, it is useful to make sure that a high enough target value is chosen so that retransmissions over the air are minimized.

Flow and Congestion Control

On the Internet, there are hundreds of hosts in each subnet, all of which are trying to get their data through to its destination. This chaotic setup can only work if each computer acts with a certain amount of manners and follows a common set of rules. When we analyze this way of designing a network, the biggest potential problem is if everybody tries to send massive amounts of data at the same time. This phenomenon is commonly called congestion (when everybody sends and no one gets anything through). To get around this problem, TCP takes note if it receives indications that packets have been lost or severely delayed (signs of congestion). TCP reacts by retransmitting missing data and simultaneously invoking congestion control (both proactive and reactive).

In order to avoid getting into a congested situation, the sender and receiver both keep a dialogue regarding how much to send. This dialogue is the size of the sending window. Factors that limit this throughput are the current load of the link and the available space in the receiver's buffer. The TCP header has a field in which the receiver can suggest a window size. The sender uses the window that the receiver suggests (after the receiver has considered the available buffer space) in order to estimate the maximum throughput of the link and a corresponding sending window. As long as the sender keeps the window below that maximum, it knows that the receiver's buffer can handle it.

In addition, the sender maintains another window, the congestion window, which varies as the load on the network changes. The sending window makes sure that the capacity of the receiver's buffer is not exceeded while the congestion window handles limitations on network capacity. Each of the two windows specifies how many bytes the sender can transmit before data is acknowledged. The number of bytes sent is the minimum of the two windows (the minimum of link capacity and buffer space).

At the beginning of a transmission, the congestion window size is set to one (one multiplied by the maximum packet size). The sender then waits for this packet to be acknowledged before sending another one. As the acknowledgment arrives, the congestion window is increased to two (and consequently, two packets are sent). As acknowledgements for each of the two packets arrive, the congestion window increases by one, giving a window size of four. This process is called the slow start algorithm and is mandatory in all TCP implementations. As we can see in this simple example (Figure 6.6), the algorithm actually leads to an exponential increase in congestion window size, and you might wonder why we call it "slow." Because a typical Internet host often

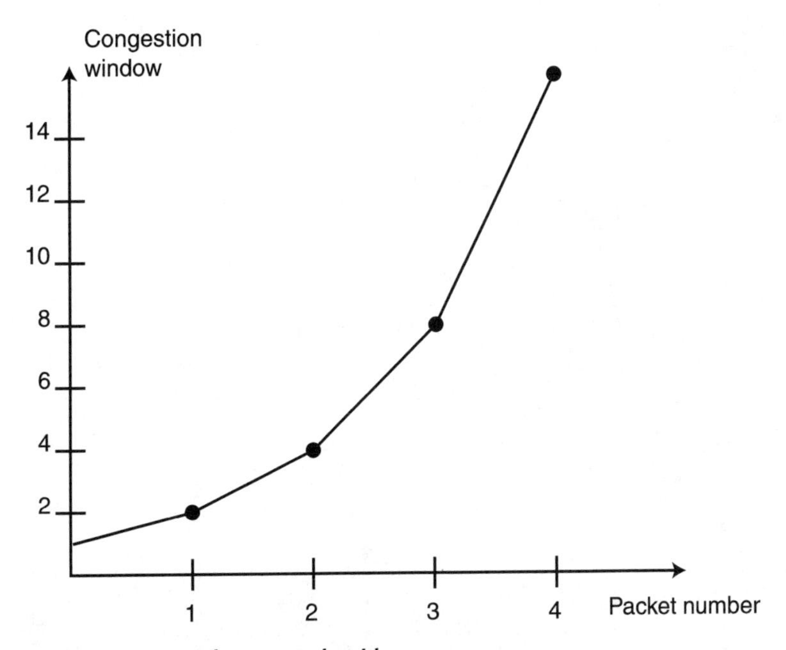

Figure 6.6 TCP slow start algorithm.

has a pretty big buffer (and the receiver will recommend a large sending window), the window that the sender uses is mostly the congestion window during the slow start. If the receiver says that it can handle 100KB, it will take seven iterations like the previous example before the congestion window will be bigger than the sending window and the sending window can be used.

Now, let's see how TCP acts when congestion occurs. For this purpose, another parameter is used, called the threshold. The threshold is like a speed limit: When the window is higher than the speed limit, it uses a linear increase instead of an exponential increase. This linear-increase algorithm is called congestion avoidance, and thus it replaces the slow start algorithm when there is a risk for congestion. In Figure 6.7, we can see how the slow start is replaced by congestion avoidance as transmission number five hits the threshold.

There are two ways in which it can detect that something is wrong: timeouts and double acks. If the sender hits a timeout for a packet, TCP realizes that it has to back off and slow down. It enters the slow start phase again, cutting the window to one and thus dramatically lowering the speed of the transmission. At the same time, the threshold is set to half the value of the window at the timeout. We can see this illustration in the figure as the new threshold value is set to 20. With the speed limit analogy again, we had an accident while going 36, so maybe 18 (half) is a better speed limit than 32. The slow start goes on and creates an exponential increase in window size until the threshold is hit

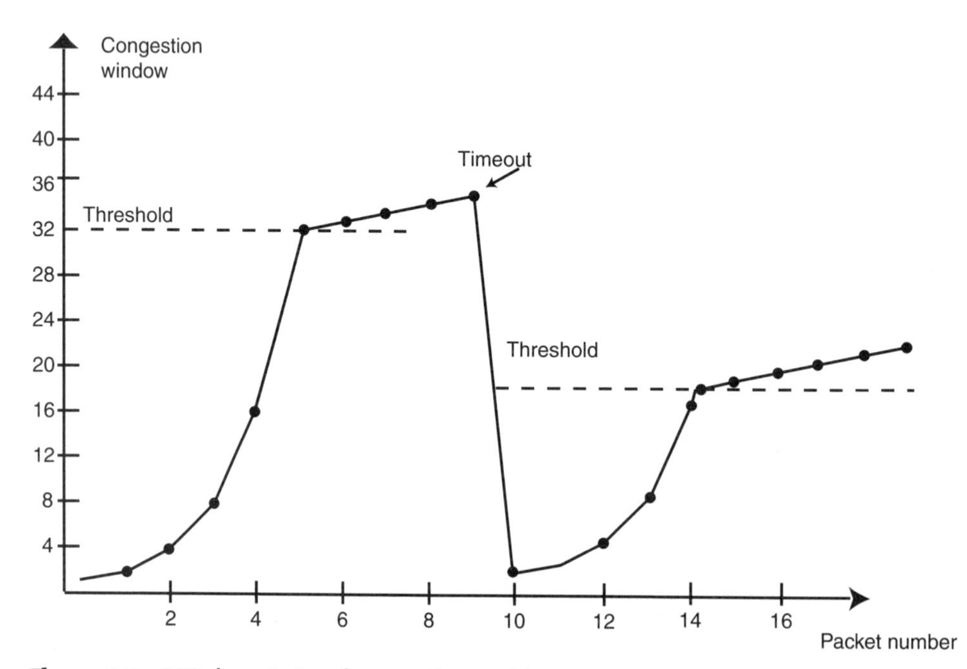

Figure 6.7 TCP slow start and congestion avoidance.

(switches to congestion avoidance and linear increase) or until a timeout occurs (enters slow start again). A double acknowledgement is less seen as a serious sign of congestion now; thus, a slow start is not needed and the window is only cut in half (entering congestion avoidance).

Developers initially made the flow and congestion control mechanisms for the fixed Internet, where lost packets indicate that the network is congested. On a wireless network, however, the delay and packet loss vary greatly and sometimes become high. The polite TCP implementations will think that the wireless network is congested and back off, therefore reducing the throughput. This situation creates the undesired result that no one gets much data through. A user who sends 20 packets per second and loses 20 percent will only receive 16 packets per second. Now, if TCP halves the throughput, the result is that only eight packets per second are successful. Therefore, running TCP directly on a highly unreliable medium, such as on a wireless network, is a bad idea. One suggestion is to use different TCP connections for the radio link and for the rest of the route to the host on the network side. The advantage would then be that the TCP connection over the radio link could be optimized for the high loss and the timeouts could be set accordingly. Apart from the fact that this method is not in line with the original thoughts behind TCP, the main drawback is that the TCP sessions would then have to be terminated at the base station. This protocol translation takes time and would hamper performance. Making the retransmissions at a lower layer of the stack is a better idea.

With the use of another protocol (on the link layer) that operates on the wireless link and covers losses, things become better. The RLC protocol is commonly used for this purpose, and it uses similar acknowledgement techniques as TCP but without the problematic flow control. Lost packets over the radio link are quickly retransmitted without TCP knowing it. The only problem now is that a lost packet is replaced with a delay, which also might cause a TCP timeout. On the fixed Internet, it is not difficult for TCP to estimate the RTT, because connections are rather predicable. As the latency changes frequently on the wireless link, however (due to the mentioned retransmissions and the load of the cell in question), it is hard for TCP to maintain a good estimate of the RTT and consequently tune the timeout correctly. The timeout of TCP is usually set in the range of two multiplied by the RTT. So, not only is the user handed a low bit rate link, but also he or she will find that the protocol used will make it feel even slower.

Hypertext Transfer Protocol (HTTP)

HTTP is the protocol that handles the transfer of Web pages and that relies on TCP (although other protocols could be used) on the transport layer for the reliable, in-order delivery of individual packets. HTTP is a stateless protocol, and it has a number of different requests that a client can use for a server.

GET is undoubtedly the most common command, and you use it every time you click a link in your browser. **GET** fetches an object from the server so that the Web browser can display the resulting page or image. Before the request is made, however, a TCP connection to port 80 of the server is set up. Next, we will go into more detail about how HTTP uses TCP for transporting the information.

PUT is not as common as **GET** and is used for uploading data to the server. **PUT** replaces the data on the server with new content.

POST is similar to **PUT**, but instead of replacing the current content on the server, it appends it. This function was originally meant as a generic append command that could be used for various applications (such as newsgroups) where data is constantly inputted into the server computer. Today, **PUT** is commonly used when inputting data to a servlet (a server-side application). As an example, you might have a Java application that converts currencies on the server that the user calls by clicking a hyperlink in a browser. This hyperlink can then be a **POST** command to the servlet, saying that the user wants to convert £20 (English currency) to U.S. dollars. The servlet makes the computation and creates a Web page with the content that is sent as a response to the client. The client browser can then show the resulting page.

HEAD returns the header of a Web page but not the message body, which is mostly useful for debugging and when you want to check to see whether a URL is still valid.

There are a few more request types that are not interesting in this context and that we will not describe further. For the definite source on HTTP, use the HTTP 1.1 specification, RFC2616.

Each request is sent as regular ASCII text, and the server responds with an object. The object that is returned can either be a simple Web page or a *Multi-Purpose Internet Mail Extensions* (MIME) response containing several objects (such as images). MIME multipart messages are defined in RFC2045 and make it possible to fetch an HTML/*Wireless Markup Language* (WML) page with all of its attached images in one single request. When fetching each object of a Web page separately, a new HTTP request has to be performed for every one. Because of the high latency of those networks, fetching each part of a Web page with a separate request might cause significant delays. With MIME multipart messages, on the other hand, several objects can be packaged into one and fetched with a single HTTP request (see Figure 6.8).

While this situation might not matter much when you are using a desktop computer that has a broad-band connection, it is a significant advantage when you are developing for wireless networks (where delays can be a serious issue, as

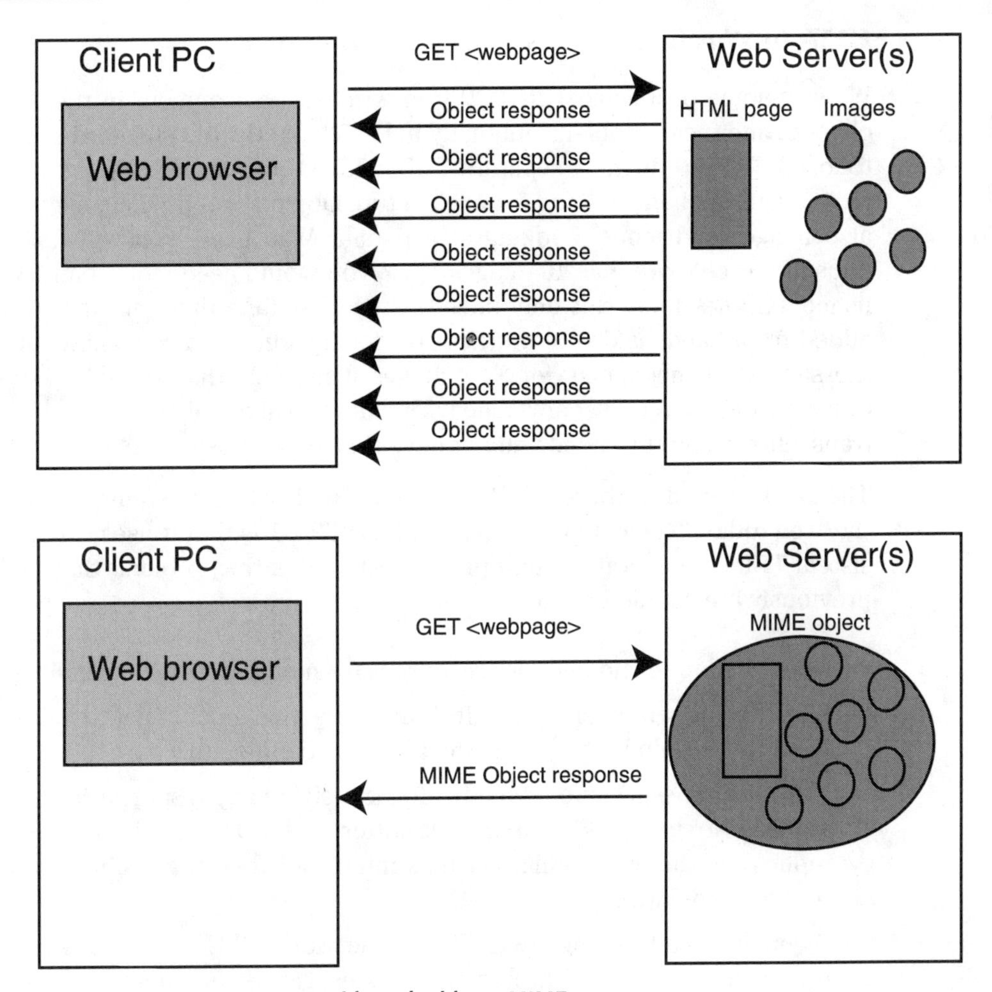

Figure 6.8 HTTP requests with and without MIME.

we will describe in depth in Chapter 8, "Adapting for Wireless Challenges"). The only drawback with using MIME multipart messages is that the caching function of the client will not help much. Say that we are fetching an HTML page with seven images, and these images are fairly static. When we package these eight objects into one, all of them have to be fetched every time if the page changes. If they had been fetched separately, some of them might have been cached and the request would not have to be sent. Fetching them separately, however, requires eight HTTP requests and a considerably higher delay. As always, it comes down to tradeoffs on a case-by-case basis.

Because the *Wireless Access Protocol* (WAP) uses HTTP requests on the server side, this description of MIME multipart messages also holds true for WAP. We discuss WAP in more detail in the next chapter.

HTTP in Wireless

We saw some problems with TCP in a wireless environment in the previous chapter, and these problems intensify in HTTP/1.0 (the old standard of the protocol). HTTP/1.0 opens a separate TCP connection for each object on a Web page, requiring handshaking for each setup. Objects can be the page itself but also images and icons. Looking at a regular Web page, you will not find it unusual to see more than 10 images; thus, you would need more than 10 established TCP sessions. Not only does this process take time, but the overhead added from handshaking and other control signaling is also significant. Also, the slow start mechanism of TCP makes it unlikely that any of these small objects will be fetched at any speed. The image will simply be fetched in a few transactions, and the speed will not have time to increase.

The work-around to these problems is HTTP/1.1, which is commonly used on the Web today. The main difference with HTTP/1.1 is that it uses a single, persistent TCP connection for multiple requests. All of the problems that we listed previously are handled in this version:

- No wasted time and network resources for multiple TCP handshaking.
- More data is sent over the single TCP connection, enabling it to increase its window size and better utilize the available bandwidth.
- Multiple requests can be issued before receiving any response. These requests are pipelined to an output buffer so that TCP can bundle multiple replies and therefore utilize the maximum packet size (avoiding many round-trip delays).
- By opening and closing fewer TCP connections, CPU time is saved in routers and hosts (clients, servers, proxies, gateways, tunnels, or caches), and memory that is used for TCP protocol control blocks can be saved in hosts.
- Network congestion is minimized by reducing the number of packets caused by TCP opens and by allowing TCP sufficient time to determine the congestion state of the network.

HTTP/1.1 also supports the compression of HTML files when the browser indicates that it can decompress the data.

HTTP/1.1 is a huge improvement compared to its predecessor. Clearly, on a wireless network, the throughput is at least doubled and the number of packets that are sent is reduced by 60 percent when using HTTP/1.1 instead of HTTP/1.0. Nevertheless, the dependence on TCP and its handshaking still makes it inefficient for short transactions.

File Transfer Protocol (FTP)

FTP is just like HTTP—an application-layer protocol. FTP is a connection-oriented file transfer between two hosts, using TCP as a transport. The only issue with FTP when running over wireless is its greed. A client that fetches a file via FTP will start one TCP session and keep it throughout the transmission. TCP's characteristics (see Figure 6.7) will make the speed of this transmission as high as possible as long as timeouts do not occur. At times when the radio conditions are good and over-the-air loss is low, one single FTP user can eat up lots of bandwidth.

As an example, consider a GPRS mobile of class 8, with four time slots in the downlink, that fetches a 2MB file by using FTP. If there are four time slots available for all of the 20 GPRS users in the cell (on that transceiver), this FTP client will constantly try to increase throughput as much as it can until it loses a packet. Theoretically, this process could cause problems if other users experience severely degraded capacity due to one user. In practice, it is likely that the wireless link will cause TCP to time out occasionally (and thus will remove some of this issue).

Solutions

So, what can the application developer do about these existing protocols that were not really designed for wireless? Well, for some it might be worthwhile to skip one or two of the protocols mentioned previously and implement the functionality from scratch or substitute them for others (we describe WAP in the next chapter). You will find it most useful to go outside these protocols if you are stuck with a rigid protocol stack that will not enable you to adjust the parameters that you want. You want to make sure that the application does not chat excessively, which always leads to increased overhead and a higher cost for the user. Trying to optimize the application on a detail level that takes into account specific characteristics of TCP is not recommended, because it has proven very difficult to achieve results that are better in all cases. Just limiting the amount of started sessions and adding a customized level of persistency can make significant improvements. This persistency ensures that TCP pushes data through more aggressively during bad connections. Here, you must use caution because the increased persistency also will result in more sent packets (which, in turn, will be more expensive for the users once packet data systems such as GPRS are deployed).

While some applications enjoy the flexibility of choosing how the protocol stack is designed, this choice is not a luxury that everyone can enjoy. First, it takes a lot of work and skills to create the communication protocols that you need. Second, some operating systems and applications platforms just will not permit it. If your application is Web based and your laptop users cannot use WAP, you do not have much freedom of movement. In Chapter 8, "Adapting for Wireless Challenges," we will look at some easy ways in which you can add other components that perform this job for you (ending up with a few middleware solutions).

Changing the Protocols

The WAP protocol stack has addressed many of these issues because it was optimized for wireless from the beginning. WAP is, as you probably know, so much more than just WML. The Wireless Transaction Protocol (WTP) does some of the things that TCP does and avoids the mentioned flaws. Perhaps we will see an increased use of parts of the WAP stack for other applications than just WML browsing as developers see the value of Wireless Session Protocol (WSP) and WTP. A comparison between WAP and the Internet model (grabbed from the not-so-objective WAP forum, www.wapforum.org) shows that a WSP/WTP/UDP stack only gives 14 percent overhead and 7 transactions in a case when an HTTP/TCP/IP stack gives 65 percent overhead and 17 transactions. Tests in the Mobile Applications Initiative (MAI) wireless labs have shown that WAP is a lot more efficient in bringing content down to the mobile devices.

Another solution is to use some kind of middleware that takes care of some of the problems for you, as we will describe in Chapter 8, "Adapting for Wireless Challenges." All together, the most important thing is to be aware of the potential problems and adjust the application accordingly, because these protocols will not change overnight.

Summary

Although lots of content will be tailored to the mobile Internet, a substantial part of the existing Internet will also be accessed via wireless networks. The TCP/IP protocol suite was designed a long time ago and is not optimized for the properties of wireless links and consequently does not perform well. Many of the problems that are associated with the existing protocols can, however, be avoided by being aware of the issues and by making the design decisions accordingly.

CHAPTER 7

The Wireless Application Protocol (WAP)

The *Wireless Application Protocol* (WAP) has become the de facto standard for delivering and presenting information on small, wireless devices. Most people have now seen what WAP looks like and how it feels, but few know how it really works. Initially, there was a lot of hype surrounding WAP, and then there was a predictable backlash (no, WAP will not end world famine). WAP does not do everything that some fancy commercials show (replacing your desktop, offering virtual reality, and so on), but even in its early incarnations, WAP is not as bad as some pessimists indicated (WAP is not crap).

This chapter will not focus on *Wireless Markup Language* (WML) (although we will describe it briefly). Rather, we will give an extensive overview of how WAP really works and what it does to cope with the properties of wireless networks.

Background and History

As we have now seen, there are some problems with using existing Internet technologies with wireless networks. Although these protocols are constantly evolving and will be widely used in the future, a number of companies wanted improvements for wireless use quicker than what the *Internet Engineering Task Force* (IETF) could provide. In the beginning, there were few early incarnations of presentation protocols for wireless.

Nokia worked on smart messaging and the *Tagged Text Markup Language* (TTML), which it used on its Nokia Communicator. This solution was Nokia-specific, however, and only described GSM/SMS networks.

Unwired Planet (now called Phone.com/Openwave) had developed HDML, an HTML-like markup language for mobile devices. This language was not based on *Extensible Markup Language* (XML), and the protocol stack was not layered and flexible. Some U.S. operators (AT&T, Sprint, and so on) started with HDML and migrated to pure WAP in order to comply with the international de facto standard. HDML and WML are similar, but making applications that work perfectly with both is not easy. HDML will therefore be phased out gradually.

The Ericsson-developed ITTP was not a Web-based model; rather, it was a platform for phone services. ITTP was not aligned with the appropriate Internet standards and lacked many elements that were needed for a complete presentation framework for mobile devices. WAP's *Wireless Telephony Applications* (WTA) cover most of ITTP's functionality today.

All of these solutions lacked flexibility and functionality, were proprietary, and were not closely aligned with corresponding Internet standards. This situation led Ericsson, Motorola, Nokia, and Unwired Planet to join forces and work on a common standard: WAP. The standardization body formed in June 1997 and was named the WAP Forum, and other companies were invited to join. Due to the $25,000 registration fee, however, member count for the WAP forum is not as great as the Bluetooth *Special Interest Group* (SIG). As of late 2000, there were some 500 WAP Forum members. Still, all of the important players in the telecommunications and software industries are now members (including Microsoft). The standardization is performed in work groups, each having different areas of responsibility. Every year, there are a number of meetings where discussions surrounding specific issues take place. As the work proceeds, more and more time is spent coordinating efforts with the *World Wide Web Consortium* (W3C) in order to ensure compatibility between WAP and the protocols of the Internet. WAP already uses existing Internet protocols (IP, UDP, HTTP, and so on) and standards wherever there are competitive solutions available.

The WAP Forum consists of members from various companies in the industry and does not create any products itself. Instead, it licenses the technology on a royalty-free basis and drives the evolution of WAP into an even better standard. Today, it not only develops its own standards but also contributes to the work of other standardization bodies such as 3GPP, W3C, and TIA. In addition, the WAP Forum has developed a framework for the certification of WAP-compliant devices.

The first release, WAP 1.0, was not satisfactory—and most vendors waited for the WAP 1.1 version before releasing handsets (WAP 1.1 is not backward-compatible with WAP 1.0). The first handsets that used WAP 1.1 hit the market in mid-1999 with the Ericsson MC218 and the Nokia 7110. While the MC218 was the first device on the market, the Nokia 7110 quickly became the most widespread

WAP device and the device for which most developers adjusted their applications. Other products followed during 2000, and developers can now develop for a vast number of devices. In addition, the WAP 1.2.1 standard was finalized in June 2000 (the release is sometimes called WAP release June 2000), which added some new features such as **PUSH**.

Many of the features of WAP are never visible to the user but contribute heavily to the user's experience. Therefore, it is crucial for the developer to understand the entire WAP protocol stack and not just the markup language (WML). For the latest specifications, one should always refer to the WAP Forum Web site at www.wapforum.org.

Overview and Architecture

Before digging into the details of WAP, we will describe WAP at a high level (and maybe in a way that people are not used to).

WAP was designed from the beginning to be device- and bearer-independent. WAP should be as appropriate for Bluetooth and GSM as future 3G networks. You must remember that WAP is not a competitor of any of these wireless systems. WAP is a suite of several protocols that all run on top of whatever wireless network is being used. We illustrate this point in Figure 7.1.

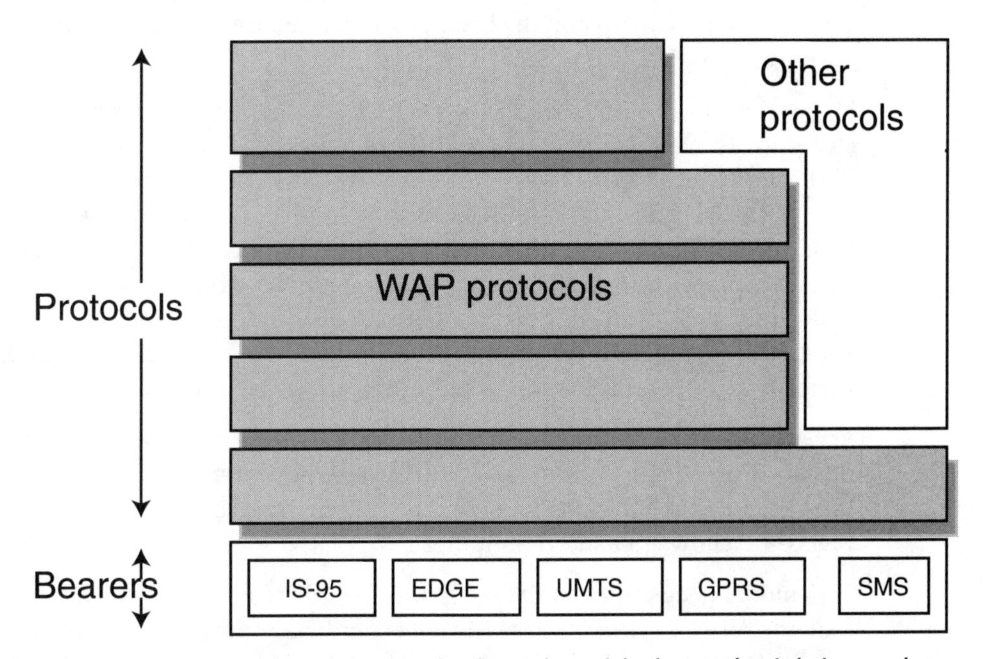

Figure 7.1 WAP is separated from and independent of the bearer that is being used.

In the figure, only the uppermost part is the WML page that the user sees. We describe the other parts in more detail in the chapter that covers the different protocols.

There are three main drivers for WAP: the market, the networks, and the devices. Looking into these drivers in more detail will lead us toward a deeper understanding of the WAP architecture.

In the mobile Internet (as opposed to the fixed Internet), the market and user behavior is different and the mobile phones and *Personal Digital Assistants* (PDAs) are not used the same way as a PC. Users need extreme ease of use and reliability, because the mobile Internet is not about surfing. The mobile Internet involves user-centric and situation-centric information that is presented so that it can be accessed straight to the point. Nobody accepts that the mobile device has to be rebooted every once in a while, but somehow we seem to accept that with our PC, the application should not put strains on the device. There is also high-price pressure on these devices, and there is a need for a low-cost solution with significant value. More and more companies are getting into the market of making the devices that we carry around and communicate with, and some are prepared to cut margins in order to gain valuable market share.

Also, the wireless networks are very different from the fixed Internet (as we saw in the previous chapter). The bit rates are generally lower, but there will also be higher latencies. As we saw in the previous chapter, this situation is not something for which the Internet was built. There will be periods of low signal quality with lots of error in the transmission. Sometimes, the user will go out of coverage and there is no connection at all.

Finally, because telephones, PDAs, and the other types of devices that WAP targets usually have small screens, little memory, and little processing power, WAP needs to address these concerns. In addition, as a technology that is aimed for the mass market, users cannot be expected to install software on their cellular telephone. Partly, this installation would make the devices vulnerable to viruses—and partly, installing applications would be complicated and undesirable by many device manufacturers. There will be other means of deploying applications on wireless devices, and we describe those methods in Chapter 11, "Operating Systems and Application Environments."

For applications that you will not keep on the device itself, you will find it beneficial to use a model like the World Wide Web, where a lightweight, generic browser accesses a multitude of applications that reside on remote servers. Figure 7.2 shows this Internet model.

This model needs to be updated slightly in order to better deal with the issues that a wireless network brings (as we described in the previous chapter). Content is placed on a WAP server, which can be a regular Web server that you configure to handle WML. This server can be located anywhere on the Internet

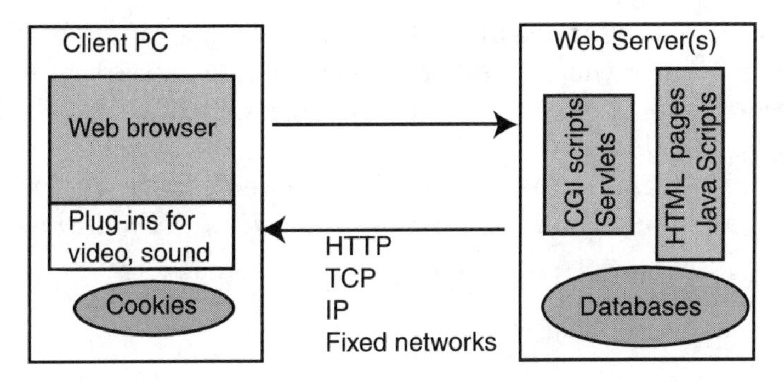

Figure 7.2 The Internet model with applications remotely accessed by client browsers.

or on the premises of the operator. The client then accesses the content with a micro browser on the mobile device. Because the route between this client and server has two very different networks over which to travel, the transmission uses different protocols for those networks. For the route that is over the fixed Internet, from the WAP server to the mobile network, the usual HTTP/TCP/IP protocols are used. Then, a WAP gateway is introduced; usually somewhere within the outer parts of the mobile network in order to facilitate the use of dedicated wireless protocols over the wireless link. Figure 7.3 shows this architecture.

The fixed part of the transmission still uses the same network infrastructure and protocols as regular Web traffic, while the wireless part now benefits from the WAP protocols that are optimized for the wireless network.

The Client and User Agents

The client usually resides in a mobile device, such as a phone or a PDA. The client is likely to contain a micro browser, which has similar functionalities as Internet Explorer or Netscape Navigator browsers for desktop PCs. We use the

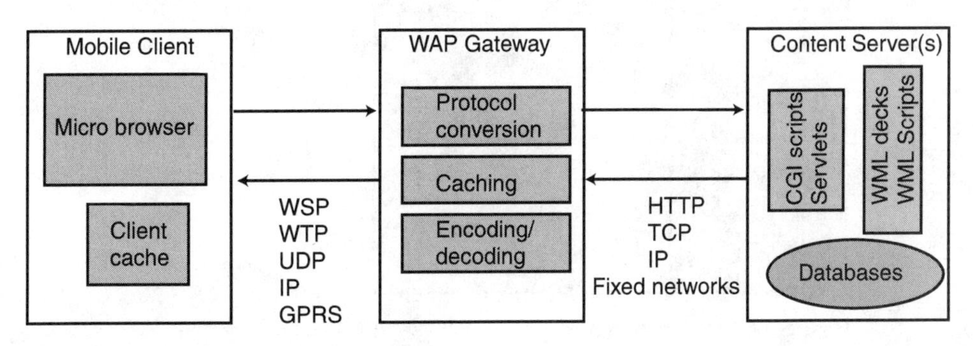

Figure 7.3 The divided WAP architecture.

term *User Agent* (UA) when we are talking about the part of the WAP device that represents the user who is communicating with the network. There might be several user agents in one WAP device, which is up to the device manufacturer to decide. The UA requests objects from the server on behalf of the user and then presents them on the device's screen. We describe this process in more detail later in the section called *Wireless Application Environment.*

Now, let's look at the heart of the WAP architecture: the WAP Gateway.

The WAP Gateway

The mobile client (UA) establishes a connection with the mobile network in question and then starts to communicate with the WAP gateway as soon as the WAP browser launches (usually by selecting Mobile Internet . . . WAP Services or similar features of the WAP device). The WAP gateway then translates the WAP protocols into regular Internet protocols (HTTP/TCP, and so on) and talks to the application server. When such a server exists, the WAP gateway also interfaces with the *Wireless Telephony Application* (WTA) server. We describe this component in more detail in the following sections. Figure 7.4 shows the internal protocols of the gateway. We describe the protocols in more detail in later sections.

The WAP gateway is actually both a gateway and a proxy (for those who are used to Internet jargon). The following tasks are performed:

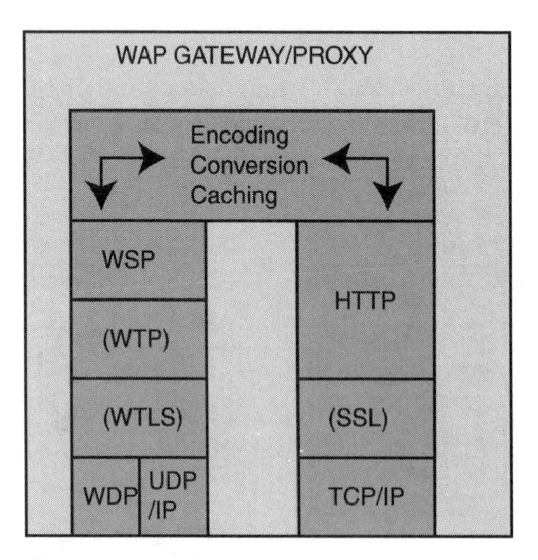

Figure 7.4 WAP gateway.

GATEWAY FUNCTIONALITY

- The gateway interfaces between the WAP protocol stack that is used on the wireless network and the Internet protocol stack that is used on the fixed Internet. As we saw previously, the right-hand side (towards the fixed network) has the same protocol stack as any gateway on the Internet.
- The gateway caches protocol headers. This feature is important in order to improve efficiency, because many protocols use similar headers throughout an entire session.

PROXY FUNCTIONALITY

- Encoding of content into binary format and compilation of WML and WML script, which are sent as text by HTTP but as binary files by WSP over the air
- Caching of content in order to take the load off the application server
- *Domain Name Server* (DNS) client that maps *Universal Resource Locators* (URLs) to destination IP addresses
- Authentication of the user, making sure that he or she has a valid subscription to WAP services

In addition, there are numerous other features that can be added to the WAP gateway. One important feature is billing support. This support enables operators (or whoever owns the gateway) to store statistics and billing information. This information can be the source ID of the subscriber who is using the WAP product, what event occurred, the date and time stamp at which the event occurred or data was logged, and an indication of which bearer the user uses to communicate with the gateway.

The WAP gateway can be implemented on just about any platform, from regular Intel-based PCs running Windows 2000/NT to dedicated, highly scalable, and robust platforms. As traffic volumes for WAP increase, you must have a WAP gateway that can handle the increase without producing more down time. Many WAP gateway vendors today have products on the market that can provide linear scaling up to a certain limit (add twice the processing power and get twice the capacity). The developer usually does not have to worry too much about what gateway the service provider/operator provides (with these important exceptions):

- The WAP gateway should be 100 percent standard-compliant.
- You should know the supported security level (if the application requires it). We describe security levels in the WTLS protocol section in this chapter.

As WAP is a fairly new standard, there have been issues regarding gateways from different vendors that interpret the standard in different ways. Wireless Test labs can help you verify your application's ability to interact with these, as described in Chapter 14, "Testing Your Application."

When setting up a configuration where the WAP browser is to access a WAP gateway, it is important to ensure that the entire communications path between the browser and gateway enables traffic on high ports to reach its destination. The WAP protocols are usually sent on ports in the 9,200 range (firewalls might have a problem with this situation). Many developers have started by developing a WAP application that they want to access on a public server. This setup is no problem as long as the developer's firewall lets this traffic through, which can be an issue, especially in larger companies that have rigorous security policies.

Some gateways also convert HTML into WML so that lazy Web administrators can take the existing Web content and provide it to mobile users. We do *not* recommend this procedure. There are, however, tools that enable the designer to keep content and presentation separate and then to tailor the presentation for the mobile audience separately. This concept is one of the core ideas behind basing WML on XML and is clearly the way to go. This procedure is completely different from taking existing content and converting it to WML automatically.

Content Server

The content server (or WAP server) can be any Web server that serves HTML pages to desktop users. We use the term *content server* here, but we can sometimes call them *applications servers*, because we will see a much wider use for these in upcoming years when applications become more diverse. The main change that has to be done to HTML Web servers is enabling WML as a MIME type. This is usually done in one of the configuration menus where the supported MIME types are listed. So, once the applications server is ready to serve WML pages and WML script to incoming HTTP requests, you have managed to configure a simple content server. As with HTML content, this server can also be used to run servlets, which are server-side applications that generate dynamic WAP content. The languages that we use are the same as for HTML content: Java, Perl, C, and so on.

When you are accessing a regular WAP deck (selecting a link), an HTTP **GET** request is sent from the client to the WAP gateway. The gateway receives the request that arrives on WAP protocols and sends it to the applications server by using a standard HTTP/TCP/IP stack. The application then sends back the requested object (for example, a WML page) to the gateway, which delivers it to the WAP client browser by using the WAP protocol stack. The **GET** request just indicates that something should be fetched from the server, and a regular Web browser also uses this command when it fetches HTML pages and other content.

With dynamic content on the applications server, we want the content that is served to the user to be different depending on some user action. If the user is converting currencies, we do not want the same page to pop up on his or her screen no matter what he or she selected, right? In order to present different content every time, a program has to be executed on the server that generates the content that is to be displayed (in some cases, we could perform this task by using WML script, as well). Those programs are compatible with the *Common Gateway Interface* (CGI), which is the same standard that is used for HTML content on existing servers. When you access one of those programs, you do not use a **GET** request because it would merely download the program to the device. The HTTP **POST** command comes into place here. A **POST** request to a program passes along some parameters to the program and executes it on the server. This program usually constructs a WML page that you can deliver to the client. This process enables complex tasks to be performed by a WAP application without consuming valuable processing power on the device. As a general rule, as much functionality as possible should be placed on the server instead of on the client, partly to consume less power but also because it is easier to make content run on multiple devices this way.

While you can surely use a generic Web server, such as Apache or Windows 2000 on standard PCs, for hosting WAP content, these servers often lack the robustness and efficiency that high-end users require. For this purpose, there are dedicated application platforms. These platforms consist of both hardware and software that when combined, achieve very high performance and robustness. We describe these in Chapter 9, "Application Architectures."

The Protocol Stack

As we mentioned previously, WAP utilizes existing Internet standards wherever possible, and the communication between the WAP gateway and the application server uses the same stack as corresponding requests on the Web to a regular Web server. There are, as we described in the previous chapter, some features missing in the existing protocols—and some characteristics do not fit the wireless world well.

HTTP is very chatty (lots of information exchanges), does not use encoding, and uses TCP/IP for transport. In addition, HTTP is purely request/response-based and cannot maintain a state with a client. Every request is treated as the first (and without the knowledge of what has been done previously), with the exception of cookies. Cookies were added as a fix to the issue of HTTP being stateless, in order for a server to drop pieces of information on the client computer. This information could then be fetched the next time that a HTTP request is sent (in order to remember passwords and user-specific configurations). There are some problems related to cookies, including the difficulty to maintain user privacy and keeping track of the total amount of cookies on the client. The latter might not be an issue on a desktop computer that has lots of *Random Access Memory* (RAM), but on a mobile phone or a PDA, every byte counts. Many HTML developers miss the cookies because they are used to having them, but there are other means for personalizing WAP content.

We thoroughly described TCP's issues over wireless in the previous chapter, and these issues were a major reason for creating WAP in the first place. TCP is also not available over all bearers, such as the *Short Message Service* (SMS) that has grown hugely popular in Europe's GSM systems. As of late 2000, there were phones that support WAP over SMS, but we have yet to see whether operators will support it. All in all, TCP was tailored for the conditions of a fixed Internet, where packet loss due to bad channel quality was rare, and it needed major changes in order to fit into the wireless world.

The WAP stack is modular and consists of an entire suite of protocols, as you can see in Figure 7.5.

Although many of the lower layers will not be very visible to users and to most developers, these layers have a large impact on WAP's performance over wireless links. Therefore, we highly recommend that developers familiarize themselves with the characteristics and features of the protocols. As we will show, some of these protocols can also be configured manually for maximum performance. The most well-known part (and the part that developers are most likely to work directly with) is the *Wireless Application Environment* (WAE).

Wireless Application Environment (WAE)

WAE is a framework that is designed to be flexible and extensible while still maintaining consistent ways of presenting information on mobile devices. We stress that the applications developer should be in total control of how the content is presented to the user. The markup language that is traditionally used, HTML, is forgiving towards mistakes by the content designer when interpreted by the browser. The opposite is true for WAP's markup language, WML.

Wireless Markup Language (WML)

Because the WAP gateway compiles WML before delivering it to the client, we need to design it according to strict rules. These rules are derived from XML.

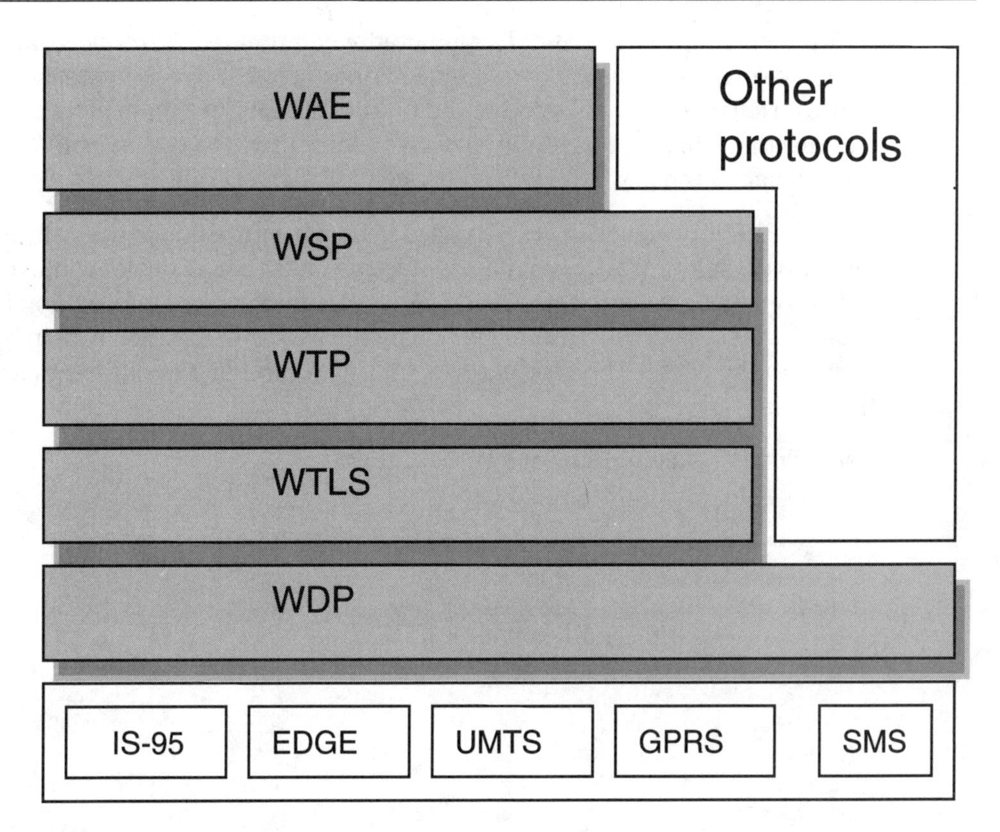

Figure 7.5 WAP stack.

XML in itself is not really a markup language. Rather, it is a meta-language—a framework and a set of rules for presentation that you can use to create other markup languages. The idea is that it should be possible to separate content and presentation so that they can evolve separately. If you add a new city to the mobile city guide, the developer should be able to add that information to a database and rest assured that the presentation will follow the same rules as previous cities. HTML 4.x, as used by many browsers of today, is not based on XML and does not have good support for this kind of separation. As we will see at the end of this chapter, an HTML version based upon XML is currently being developed (called XHTML).

While HTML is interpreted as a top-down description of the visual contents of the screen, WML also has the properties of a programming language (such as variables and timers). This functionality is made possible through WAP's support for states (that the same session can be resumed as a user returns from being out of coverage or on another bearer). We describe this concept in more detail when we describe WSP in the following sections. Web designers who are accustomed to working with 21-inch desktop computers and the richness of

HTML content often become disappointed when they are introduced to WML on a five-row phone display. WML indeed does not yet have the extensive support for rich media that is often found on the Internet, but there are numerous features that will prove valuable as developers get used to them. Table 7.1 illustrates some of the key differences between HTML and WML.

While HTML is designed in HTML pages (a flat structure that is basically a document with links), WML uses a deck-of-cards metaphor. A deck is pulled from the server with a transaction (see the WTP section for more information about transactions), is encoded into bytecode by the server, and is then sent in binary format to the client. The user can then navigate through the different cards in this deck.

The following example shows a WML 1.1 deck:

```
<?xml version="1.0"?>
<!DOCTYPE wml Public "-//WAPFORUM//DTD WML 1.1//EN>
<http://www.wapforum.org/DTD/wml_1.1.xml>
<wml>
<card id="first" title="My deck">
<p><b>A non-Poker WML card</b></p>
<do type="accept">
<go href="#second"/>
</do>
</card>
<card id="second" title="Second card">
```

Table 7.1 Differences between HTML and WML

HTML 4.X	WML OF WAP 1.2.1 AND BELOW
Content in HTML pages	Content in a deck-of-cards metaphor
Extensive support for layout and design, such as frames	Text and images
Various media clips, such as music and video	Not supported
Color images and animated graphics	One-bit graphics (black and white)
Links and server-side applications	Similar support, using HTTP 1.1
No variables or states	Variables and state supported, even when returning from being suspended (out of coverage, and so on)
No programmable shortcut keys	Shortcut keys supported
Not based on XML	Based on XML
Cookies available for session control	No support for cookies
No events or timers	Events and timers supported

```
<p>Another interesting card</p>
</card>
</wml>
```

This example presents a first card that enables the user to select a link to a second card. More than likely, most people will not write WML code directly; rather, they will use an editor such as Microsoft FrontPage or Dreamweaver. *Software development kits* (SDKs) are also available from the major device manufacturers: Nokia, Ericsson, and Motorola (to name the three biggest). We include some SDKs on the CD-ROM that accompanies this book, and more are available on the book's Web site. Because WML is based on XML, there will be less of a focus on the actual WML code and more emphasis on managing the content (remember that content and presentation now can and should be kept separated).

Although you can use variables and events in WML, it is still a pretty static way of presenting information. Just like dynamic HTML can be generated through server side programs (servlets), dynamic WML can be used to create customized WML decks for mobile users. Depending on what the user selects in a menu, a different WML deck can be generated by the server and sent to the client. We describe the session-layer view of this process later in the *WSP* section. For further reading, *WAP Servlets* by John Cook (Wiley, 2001) provides a good guide for getting started. We will not explore WML and WML script and design here, because there are entire books that focus on that topic. The SDKs on the CD-ROM contain some examples that are useful for getting started.

WML supports images, but these images are still just one bit per pixel (and therefore, black and white). The images can either be stored on the application server as regular bitmap (.bmp) images or as *Wireless Bitmaps* (WBMPs). We usually recommend using the bitmap format and letting the WAP gateway handle the conversion (.bmp to .wbmp). While this procedure might sound inefficient, it does not require much time and power from the gateway to be noticeable. The advantage is that you will not have to redo your images if the .wbmp format is changed in later revisions of the standard.

As we saw in the previous architecture description, WAE uses many of the features of existing Internet standards (such as HTTP and IP). In addition, URLs are still used to access application servers (for example, wap.picofun.com) and the languages that we have defined so far. WML and WML scripts are derived from XML and ECMA scripts (Java scripts).

You might now wonder how the *Handheld Device Markup Language* (HDML) fits into this picture. HDML was a predecessor to WML, which originated in the labs of Unwired Planet (Phone.com/Openwave), which was not based on XML. HDML can still be found in old implementations of mobile Internet services, but everyone (including Phone.com/Openwave) is now striving toward using XML-based content (in other words, WML and XHTML) for all new services. As always, we strongly advise you to follow standards strictly when they are available.

The last part of WAE is the support for integration of telephony functionality, which the *Wireless Telephony Application* (WTA) handles.

Wireless Telephony Application (WTA)

In some mobile devices (such as a smartphone), there is a software part (operating system and applications) and a telephone part (including hardware and software). As these mobile devices gain more and more applications functionality, we want them to also have the features of a telephone, such as making calls and storing numbers in the phone book. WTA is designed to provide this functionality through a standardized *Wireless Telephony Applications Interface* (WTAI). An example application is the WAP-enabled Yellow Pages application that enables you to find a nearby restaurant. As you obtain a number of listings on your device, you want to dial the restaurant in order to check availability (while some restaurants might let you book your reservation by using WAP). Instead of writing down the number and then dialing, WTA enables you to click the link in order to make the call. In another case, you might want to add the number to your phone book by clicking it.

We show the architecture for WTA in Figure 7.6.

As we see in this figure, WTA introduces a new node to the WAP architecture: the WTA server (or WTA origin server, because this server is where WTA applications originate). Because it is very important to protect the user's integrity and security, the WTA server is connected to the WAP gateway via a dedicated connection (most commonly placed within the operator domain).

There are three categories of WTA features:

- Network Common WTAs, which are common to all bearers and are available on all networks
- Network Specific WTAs, which includes features that are only found in certain networks or that are provided by specific operators
- Public WTAs, which are simple features that can be accessed in third-party applications through the standard WAE

Public WTAs are likely to be the most frequently used, because they are the easiest to access for developers. Two such features are implemented in WAP 1.2.1: Make call and Send DTMF tones. Make call is exactly the feature that we illustrated in the previous example, where a user can call a number by simply selecting it in the WAP browser. DTMF tones are the tones that you generate as

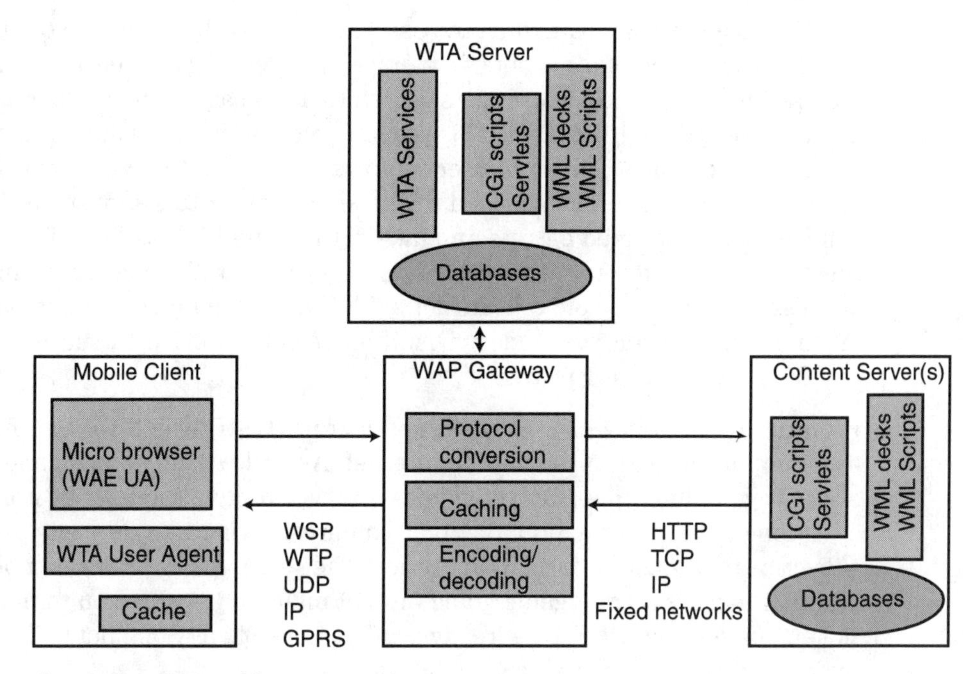

Figure 7.6 WTA architecture.

you dial a phone number (might or might not be data-enabled) The Send DTMF tones imitate the feature that many call centers use today where you press a number to navigate a set of menu options.

Wireless Session Protocol (WSP)

Wireless Session Protocol (WSP) brings some of the key differences from the TCP/IP suite to the WAP table by introducing sessions and connections that have a client state. With the foreseen impact of personalized services of the mobile Internet, it was vital for the server/gateway end to keep a more lasting relationship with the client than what HTTP offers. The HTTP request/responses are forgotten as soon as they have been executed, and no permanent state is kept between the client and the server. This situation makes it hard to use variables and other state indicators that personalized services require. The most important difference between this solution and the Web solution with cookies is that cookies keep information about the user and the application while WSP keeps information about the connection and its properties. The information that cookies can hold (mostly personal information) should instead be handled in a more secure way, such as storing it on the user's SIM card.

The WSP state is also kept when the user is out of reach of the network (when the phone is turned off or is out of coverage). Suspending a request quickly and then resuming it when the user gets back into action again enables parameters and the client state to be kept. This suspend-resume functionality is also used when a user changes bearers. A user could use WAP over *General Packet Radio Services* (GPRS) and then suspend the session in order to establish an HSCSD call with a higher, fixed bit rate and then resume the WAP session. During the time that the session is suspended, it does not matter if the client changes IP addresses or keeps the old one. In fact, WSP is not reliant on lower layers of the protocol stack when suspended, which enables the device to run more efficiently and save battery power.

From WAP 1.2.1 on, WSP also has support for **PUSH**, which we also foresee as an important part of the mobile Internet. As GPRS is introduced and users are always online, this feature enables a server to give the user customized, situation-centric information. Say that you are arriving in a new city and you subscribe to a soccer information service. The server can then send a notice to you that there is a great game going on that night and enable you to purchase tickets on the spot. **PUSH** is already available in GSM as SMS, but GPRS introduces this feature to any IP-enabled device.

Although WSP differs in these matters from HTTP, it is largely a binary-encoded version of HTTP/1.1. The same **GET** and **POST** commands that we use on the Web are used here, but content is sent as binary ones and zeros instead of plain text. This feature dramatically reduces the amount of data that is sent over the air and increases the efficiency of WAP.

Because WSP uses headers that are well known to both the gateway and the client, you can encode them into short, binary sequences. This procedure can shrink 32 bytes in plain text into two bytes of encoded representation. This shrinkage saves important capacity over the wireless link, and as we will discuss in the next chapter, it probably also saves the users some money. This dramatic gain in capacity is not only due to the non-binary representation of HTTP headers but also results from avoiding the repeated static headers that HTTP uses. WSP reduces the amount of redundant information that is sent and minimizes overhead. When several packets of the same session are exchanged, WSP can also cache the used headers so that they are not sent to the user every time.

WSP offers the application two connection types: connection mode and connectionless. Both have their advantages and disadvantages, and it is up to the application to choose. Newer WAP browsers provide the option to choose between the two, which increases the developer's feeling for WAP's characteristics (to play around and switch between the two). The difference in handling interruptions is especially interesting, as we will see in the next chapter.

Connectionless WSP was the most commonly supported service in the first WAP phones because it is easier to implement. Connectionless WSP uses the same send-and-forget philosophy as UDP, where packets are not acknowledged. In fact, when running connectionless WSP over an IP bearer, UDP is used as a datagram service. Because there is no established session, more header information has to be included in each packet in order to route it correctly (which leads to higher protocol overhead).

Connection mode WSP, on the other hand, offers this session management and also provides reliable transport with acknowledged packets. The session starts with a negotiation of service capabilities between the client and the server—parameters that are expected to be valid throughout the session. Both the client and the server can abort the session. The regular **GET** operations (getting objects from the server, such as WML pages) can be made reliable, but **PUSH** can also be performed as either confirmed or unconfirmed. With confirmed **PUSH**, the client sends a receipt back to the originating server and the client can also choose to abort a **PUSH**. Note that WAP **PUSH** was *not* supported until WAP 1.2.1. We show confirmed **PUSH** in Figure 7.7.

Note that a client cannot receive **PUSH** when in a suspended state.

The reliability that the WSP connection mode offers is supported by the *Wireless Transaction Protocol* (WTP), which ensures that the packets are delivered.

Wireless Transaction Protocol (WTP)

WTP was designed to become something that is more reliable than UDP but not as heavy as TCP. WTP is essentially a lightweight transaction protocol that hides bad network conditions from the upper layers of the stack and from the user.

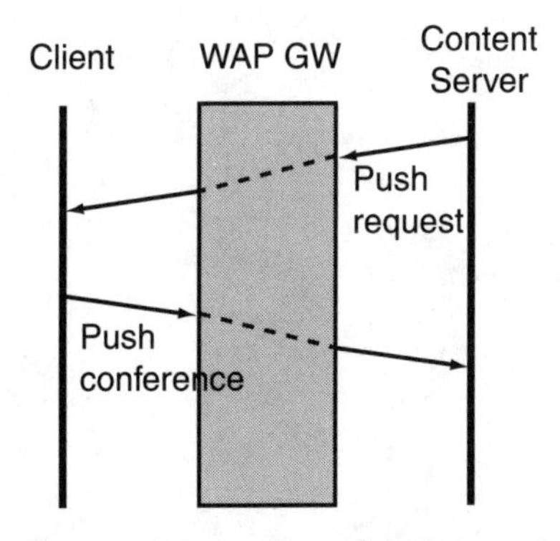

Figure 7.7 Confirmed PUSH.

Although this description sounds like a TCP version that is optimized for wireless, WTP actually also handles some HTTP functionalities. While TCP is packet-oriented and makes sure that the packets are delivered reliably and in order, WTP is message-oriented and delivers messages rather than a stream of packets. In the HTTP/TCP case, HTTP works on the message/request level while TCP only looks at the packet/data segment level. You can compare this situation to delivering an entire truck of moving boxes instead of focusing on delivering individual boxes that someone else assembles at the destination. This model is more natural for banks, for instance, which are message/transaction-oriented. The WAP model is, therefore, more intuitive for mobile banking applications than for the corresponding fixed-Internet counterparts. The protocol tries to minimize the number of transactions due to the delays of wireless links.

Unlike TCP, which uses a three-way connection, WTP does not have either connection setup or teardown. There are three different classes of transaction services:

WTP Class 0. Unreliable invoke message with no result message (unreliable **PUSH**).

WTP Class 1. Reliable invoke message with no result message (confirmed **PUSH**).

WTP Class 2. Reliable invoke message with one reliable result message (regular, reliable message transfer).

Figure 7.8 shows a WTP Class 2 transmission, which could be a client requesting a URL from a WML deck. The responder formulates a message, which means that it fetches the deck from the content server.

If it takes too long to process the received data, there could be a risk that the initiator has timed out and that it will resend the message. To avoid this situation, the responder has a similar timeout—after which it sends a hold-on message

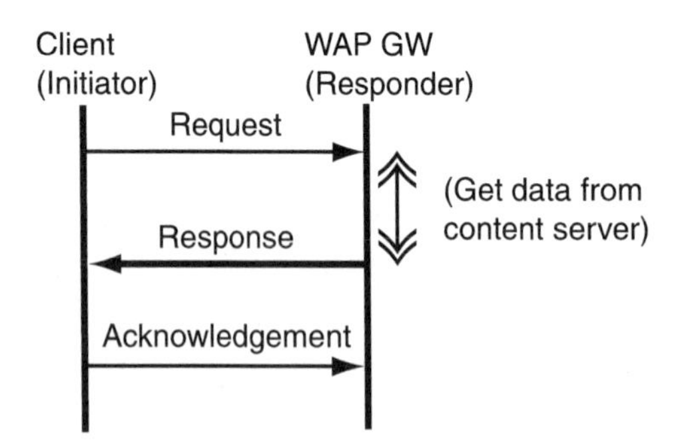

Figure 7.8 WTP Class 2 transaction.

telling the initiator not to resend the message. If User acknowledgments are enabled, a separate acknowledgment is sent when the message is received (separate and before the response).

WTP was, of course, designed with WSP's needs in mind, and this design is the primary purpose of the protocol. In the future, however, there will be no stopping you from implementing your own application on top of WTP. This implementation requires the target WAP devices to support this flexible use of the stack. At least the higher-end devices that have open platforms are likely to support this functionality.

WTP does not use any security mechanisms, but you can optionally implement them by using the *Wireless Transport Layer Security* (WTLS) Protocol.

Wireless Transport Layer Security (WTLS)

In the push to reuse Internet standards where applicable, developers based WTLS on the *Transport Layer Security* (TLS) protocol, which was formerly known as the *Secure Socket Layer* (SSL) protocol. Because TLS can rely on TCP for some reliability functionality while WTLS cannot, WTLS covers this part itself. In other words, WTLS can operate over UDP and TLS/SSL cannot. The WTLS layer is modular, and the application that is used decides what security level and features it will use. All in all, the aim is to provide data integrity, privacy, and authentication between two communicating applications. Like all protocols in the WAP stack, WTLS is optimized for bearers that have relatively low bandwidth and high latency, but it still has some issues that should caution developers from using it too much.

WTLS appears much like TLS to the application and end user. In order to start a secure session, the following

```
HTTP://wap.whatever.com
```

is replaced by

```
HTTPS://wap.whatever.com
```

At the start of a secure session, a handshake procedure is used where parameter negotiation takes place. At a high level, the client proposes a set of parameters and a security level. The server then responds with a revised proposal, which *cannot* lead to a higher level of security than the client originally proposed. During this process, the client and the gateway are authenticated—and

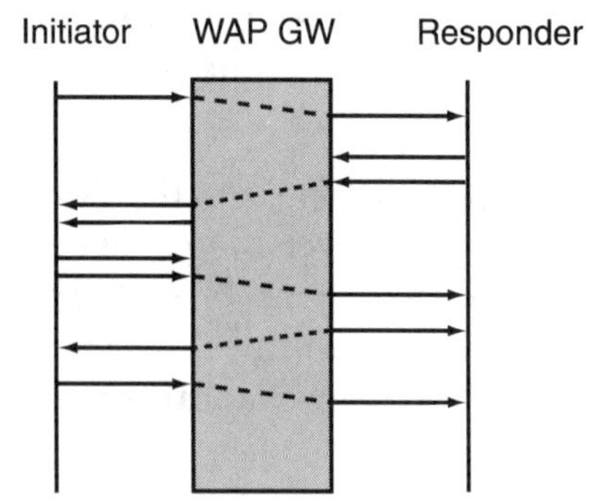

Figure 7.9 WTLS handshake.

it is decided whether or not the data should be compressed or encrypted. If certificates are to be used, this decision is also made during the handshake (when the WTLS security class is selected). Figure 7.9 shows that this handshake requires a substantial number of transactions.

Because WTP does not support the kind of handshakes that TCP uses for setup, WTLS has to include it when security is needed. This handshake requires a substantial number of transactions; therefore, an optimized version has been defined, but it still takes a lot of time to complete. In high-latency environments, this situation can be a real problem—and it remains to be seen how users tolerate this extra setup time.

The following security classes can be negotiated during the handshake:

- WTLS Class 1: No certificates
- WTLS Class 2: The server has a certificate, but the client does not. This class is expected to be the most commonly used class.
- WTLS Class 3: Both the server and the client have certificates.

Another problem with WTLS is that the certificates can produce a substantial overhead for the session. A typical certificate is around 1KB, which is about the maximum deck size for some WAP devices. If WTLS sessions are short, this amount can be a large part of the entire amount of data that is transferred. Therefore, you should only use the security features of WTLS if absolutely necessary.

The device's specifications should mention which WTLS class is supported.

Whether or not WTP uses WTLS, a datagram service is still needed to get the messages sent over the networks. For IP networks, regular UDP over IP can be used, but for bearers such as SMS, an adaptation protocol is needed: the *Wireless Datagram Protocol* (WDP).

Wireless Datagram Protocol (WDP)

The transport layer in WAP is divided between WTP and WDP. WDP serves as an interface between whatever network bearer is used and the higher-layer protocols. WDP is intended to hide differences of running over, for instance, a GPRS network with IP and an SMS over a GSM network. In order to reuse components from the existing Internet standards, UDP is used whenever the bearer is capable of running IP.

UDP is merely a delivery protocol and does not resend lost or delayed packets. For the fixed Internet, UDP is commonly used for real-time applications where it is pointless to receive delayed information. These instances include streaming music and video as well as online games such as Quake. This kind of application is perfectly suited for UDP's send-and-forget philosophy. In WAP, this datagram service is complemented by higher-layer functionalities, such as the retransmission of lost packets. Segmentation and reassembly are handled by IP, which means that WAP in itself might not restrict the size of packets. TCP measures the *Maximum Transmission Unit* (MTU) and uses it throughout a session. This functionality is not included in WAP; consequently, IP will segment its packet when it reaches a network that has a low MTU. Some network elements that analyze IP packets might encounter problems with analyzing these segmented IP packets. This analysis will then only show when the WAP decks that are delivered are larger than the MTU of the narrow part of the network. An example is a 2,000-byte deck that is sent over a network that has an MTU of 1,500 bytes (Ethernet). IP will then be forced to split this message into two packets, introducing IP fragmentation. The network nodes on the wireless side of the WAP gateway then need to be able to handle this situation. Fortunately, the regular developer does not need to worry about this situation because it happens mostly in the operator's domain.

WDP also handles the port number for the application, enabling several applications to run concurrently on the same device.

When the network bearer (such as SMS) does not support IP, WDP introduces an adaptation layer that hides this information from the upper layers. Figure 7.10 shows how different bearers need different amounts of adaptation. The more advanced the service that the bearer offers, the less adaptation is needed. Bearers in the rightmost part include GPRS and UMTS, which are pure, IP-enabled bearers.

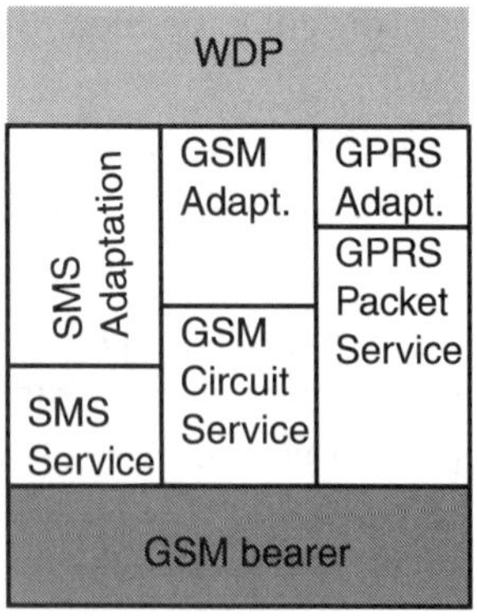

Figure 7.10 Different bearers and the level of service that they offer.

WAP Now and in the Future

As WAP services are launched in countries all over the world, some people are getting their first experiences with the mobile Internet. There is incredible activity among WML developers today, and everyone wants to jump on the bandwagon. Still, you can hear some HTML developers becoming grumpy about abandoning their favorite markup language and learning a new one that is extremely unforgiving to syntax errors. Others say that higher bit rate technologies such as GPRS and 3G will take away the need for WAP in the future. Here are some reasons why we think that WAP is here to stay and why it is likely to evolve in the future. Here, we use the traditional way of naming WAP standard releases: WAP 1.1, WAP 1.2.1, and so on as well as the time of completion format, such as WAP June 2000.

What's New in WAP 1.2.1 (June 2000)?

While the first wave of WAP applications used WAP 1.1, more and more developers are now glancing at the new releases. WAP 1.2.1 was finalized in mid-2000 and phones started to appear early 2001. The main features introduced in 1.2.1 are as follows:

PUSH. This feature enables trusted applications servers to initiate the transmission of information to WAP devices. This is especially useful for networks such as GPRS, where users are always online.

UAProf. This introduces user agent profiles that you can use to advertise handset capabilities to applications servers and to other entities. In other words, you can better adapt content to different types of devices.

Wireless Identity Module. Enables storage of user data on the handset. The SIM module used in GSM/GPRS phones can then be used as WIM.

WAP 1.2.1 also includes a number of smaller adjustments that take away ambiguities. For a comprehensive discussion of these updates and others, check out the WAP Forum Web site at www.wapforum.org.

After WAP 1.2.1, the next phase will contain fewer major changes and more adjustments and corrections. Both WAP 1.2.1 and its successor will be backwards-compliant with WAP 1.1 to ensure that new phones work with legacy content. Also some security add-ons, like Wireless Public Key Infrastructure (WPKI) support, are likely.

WAP Next Generation (WAP-NG)

Then, the next step has the working title WAP-NG and will be a larger step. We view WAP NG as the standard that initial 3G handsets will use, and lots of work has been put into it already. NTT DoCoMo is heavily involved in this development in order to find a solution that can be used in its I-Mode service as well. Today, the massively popular Japanese service uses a proprietary HTML derivative called c-HTML, which has caused some worries among WAP (WML) developers. The issue with HTML and HTML derivatives has so far been the lack of strict rules for content description. When designing for small devices that all have different form factors and screen sizes, you must be able to control the way that content is presented in a more detailed way. WML is derived from the rules set of XML, a generic content description framework. Now, the *WWW Consortium* (W3C) is working on ways to apply XML rules to a revised HTML standard: XHTML. The WAP forum is working closely with W3C in this effort, and we foresee that WAP-NG will support both WML and XHTML, but the WAP Forum should be your source for the latest developments. Figure 7.11 shows the evolution of the markup languages.

This picture is not perfect, however, and we need to make some comments. XML is really a framework and a rules set from which WML and XHTML are derived. Because this version of the WAP standard is not yet finalized, these things are still not carved in stone.

Why WAP in 3G?

With bigger, better, and faster networks around the corner, one might ask why we will need WAP in the future. Then the delays should be lower and the need for optimization will not be as great, right? In addition, there will be fancy devices with big screens and more processing power. Is there enough in WAP to make it a force in the future as well?

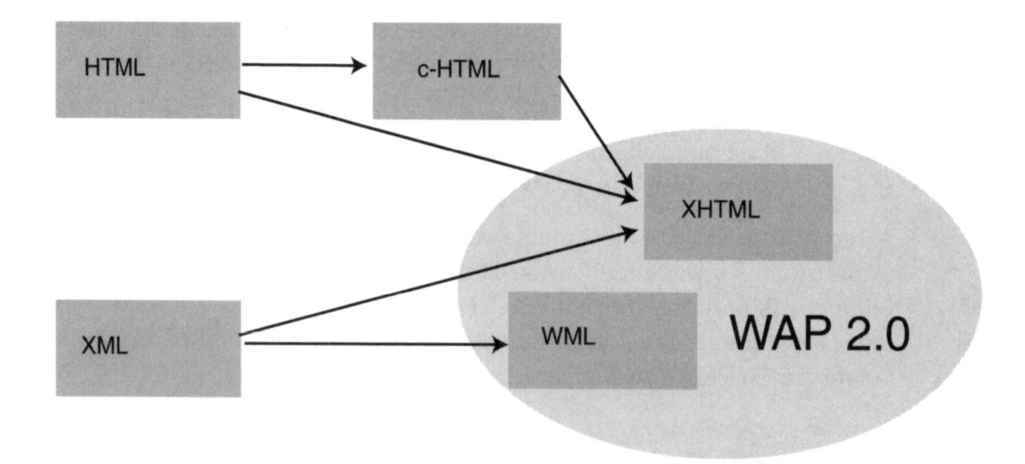

Figure 7.11 Migration of the markup languages.

One big reason for WAP's continued presence as a mass-market access is that all major phone manufacturers have committed to put WAP into all upcoming phones. Apart from that, I would like to explain four main aspects of WAP that are important now and in future 3G systems:

Bringing content down to small devices. WAP was built from scratch in order to define ways to format content for small devices that had a number of different form factors.

Efficiency. As packet data becomes the preferred way of using many mobile Internet information sites, users are likely to pay more corresponding to the amount of data that they are transferring than based on the time that they are connected. In other words, everyone is keen on having as little protocol data as possible sent for each bit of information. Using WSP's binary encoding, WAP increases the efficiency of transmissions over wireless.

Robustness. Wireless connections are likely to sometimes be disrupted by the lack of coverage or radio shadowing objects (such as elevators). Applications need to be robust and rebound after tough treatment (a feature that WAP adds). The Suspend/Resume feature ensures that the session can be kept even after long interruptions.

Telephony integration. In order to truly leverage the unique features of the mobile Internet, one has to see the synergies between the data part of the handset and the phone part. If I search for a florist and get three names, I want to click one of the names on my phone's display rather than scribble down the number and dial it. The *Wireless Telephone Application* (WTA) framework adds this feature to WAP.

Finally, it is clear that WAP is not perfect, and there will always be room for improvement. This situation is natural, however, when a standard is developed by lots of companies over just a few years. The advantage of this slow migration to a great applications environment is that content for a great diversity of devices is specified in a unified, optimized way.

There are times when you will have to go outside WAP in order to get more functionality for your application, however. You might want to deploy an application on the device or just have a downloadable Java application where you are in charge of the protocol stack. Whenever browsing and transaction functionality is needed, you should use WAP; but for other cases, Chapter 11 will give you some useful advice.

Summary

WAP is a suite of protocols rather than a single protocol. The lower layers ensure that information is sent efficiently and securely to a mobile device that has a WAP browser. The WAP gateway converts protocols that are used on the Internet to those that WAP uses over the air. While the markup language of WAP (WML) is not as easy to program as HTML, there are still some important advantages of WML, including its session maintenance, variable handling, and its relation to XML. Even as the bit rates get higher in 3G networks, there is a need for a suite of protocols that ensure efficiency, robustness, and telephony integration. The markup languages will be based on XML so that content and presentation can be clearly separated.

Adapting for Wireless Challenges

Without a doubt, as we have seen so far in this book, wireless networks are different from their fixed counterparts. Clearly, applications developers need to be aware of how the networks affect the applications. As we now enter a new era of wireless applications (with the advent of packet data and always-online devices), there are new issues that we (and even established mobile Internet applications developers) must consider. This chapter will examine some wireless properties that affect applications and describe how to overcome these challenges. This information applies to both established developers who are already experienced with wireless (who might not have dealt with the problem before or who have not designed for packet networks) and to those who come from the *Local Area Network* (LAN) and fixed-Internet world. Both kinds of developers not only need to understand how the wireless networks work (by reading Part One of this book), but also need to gain a feeling for those networks and how to create good applications for them. We should note that much of the advice that we give in this chapter applies mostly to those cases where the development uses a programming language such as C or Java, even for the client part. For thin client applications such as *Wireless Application Protocol* (WAP) applications, the developer of course has limited possibilities for affecting wireless performance, but some of the information applies to those situations, as well.

What Affects Applications and Why?

If this book is your first acquaintance with wireless networks, then all of this information probably feels overwhelming and mighty confusing. While the first step is understanding the basics of what the architecture looks like and how it all works, an even more important step is understanding how you can affect the applications and how you can optimize them. The key here is finding the disease *and* the cure. We have heard many developers complain that everyone is whining about *Transmission Control Protocol* (TCP) behavior over wireless without suggesting feasible remedies. Obtaining information about how some clever guy in a laboratory is working on a new and enhanced version of TCP will not help your application work more efficiently in today's network. Do not misunderstand this statement; there are definitely some brilliant people working on such solutions. But here, we will concentrate on what you can actually do in today's world. When we examine network properties, we do not want to plunge deeply into the details of lower-layer protocols and implementation parameters for several reasons.

First, you do not want to optimize your application for a specific network; rather, you want to make your application generally optimized for wireless networks. Many of the most important properties of wireless networks, such as interruptions and latency, are common to all networks. Many developers also discover that their efforts toward making their applications suitable for wireless networks makes their appliations work better on fixed networks, as well. If the e-mail server goes down, for example, the application that is tolerant to interruptions will not become confused and helpless, but instead will handle the problem smoothly.

Second, it is easy to get into suboptimization when you are digging too deep into protocol-specific features. The TCP properties that we discussed in Chapter 6, "Unwiring the Internet," are very valuable for a developer to understand, but tailoring the application for these properties is another story. You might be tempted to try to detect congestion before timeouts occur in order to avoid a slow start, or you might force TCP into slow-start mode in order to achieve a more rapid ramp-up of the speed. The problem is that you will never be in control of all of the factors that affect the performance, and predicting and controlling this process in detail is very difficult. Not only are there several implementations of TCP alone, but there are also many complex interactions between the different protocol layers.

Finally, you might have optimized the application in detail to work excellently with one set of parameters or with one wireless scenario, but later on the application might work even worse than before in other cases. Before coming into labs for testing, I have seen developers bombard the lab experts with questions about how to configure the lower layers of *General Packet Radio Services* (GPRS) and *third-generation* (3G) networks in the emulator. They also ask how operators will tune their networks and set parameters. I tell them that they do not want to know that information, and while it is valuable to know about what some of the *likely* settings will be, you should not optimize an application for a specific set of parameters. Different operators will use different settings, partly because their geographical topology is different, but also because other factors differ, such as interference from other networks and radio sources. Each network implementation is a totally different story, and the applications developer should not worry too much about the details.

Now that we have set the stage, let's move directly to the one property that my experience shows is the most difficult for applications to cope with: interruptions.

Handling Interruptions

In the first wave of GPRS and 3G applications, we will see both existing mobile Internet developers (WAP) and some that are porting their existing *Local Area Network* (LAN) or Internet PC applications to the wireless world. While the latter act like sitting ducks for getting into trouble with the many different wireless networks, interruptions are difficult even for those who are experienced with developing for existing circuit-switched networks. This situation is due to the differences in characteristics of packet-switched networks, such as GPRS, where users are always online.

The main difference with interruptions in these two kinds of networks is that connections can rebound in packet-switched networks. When you are using an application and walk through a tunnel for a few minutes, you cannot expect the application to work in the tunnel because the radio signals might not be capable of reaching you (although you can fix this situation, and in Sweden, cellular telephones even work on the subway). The difference occurs when you come out of the tunnel with your wireless device and take a look at the application. The circuit-switched connection was lost when you entered the tunnel, and you would have to dial the *Internet Service Provider* (ISP) again in order to restart the application. On a packet-switched network, on the other hand, the network connection will come back as you emerge from the tunnel and enter the base station's range again. The user therefore expects the application to resume where it left off. Will your application do that?

Before we study the different means of circumventing the problems of interruptions, we need to determine what makes interruptions occur (see Figure 8.1). An obvious situation is the one that we just described, where the user (for a brief period) moves through an area that radio signals cannot reach and a radio shadow develops. A tunnel is one place where this situation can occur, but elevators and basements are also typical areas where radio signals have a hard time finding you. In this case, we must not forget Bluetooth, which has a hard time penetrating metal and other compact materials. Generally, a radio signal is better at penetrating solid objects and spans a greater distance if the frequency is low. Bluetooth uses the 2.4GHz band, which is even higher than

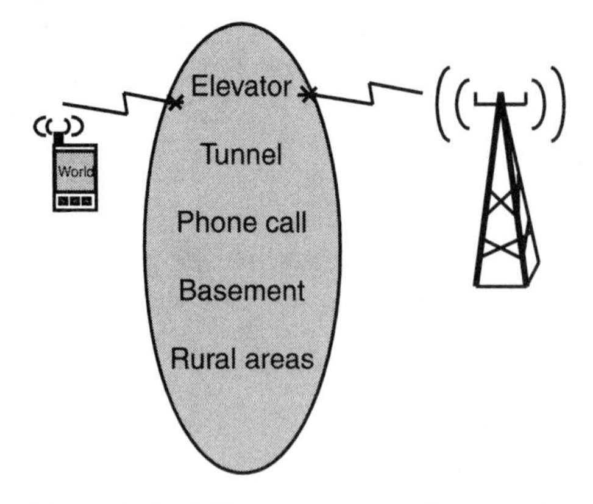

Figure 8.1 Different causes of interruptions.

the 2GHz that UMTS networks use; therefore, Bluetooth should be a bit worse at penetrating solid objects. A regular wall of an apartment is no problem, however.

Interruptions also occur when a wireless user moves out of the coverage area and signals cannot reach the base station. This location could either be a spot where the operator in question has a gap in coverage or where no wireless coverage exists (if the user traveled beyond populated areas, for example). In either case, the network is not capable of reaching the user's device for seconds, minutes, or perhaps even hours. This situation is likely to be one of the most common reasons for interruptions, especially in the initial buildout period. Luckily, most developers design networks with this factor in mind (such as GPRS, where the software upgrade of base stations makes it possible to quickly update an entire network). We predict, however, that there will still be areas of no coverage in the future.

A last example of interruptions is specific to GPRS class B terminals. Recall that you can use a GPRS class B terminal for both voice and packet data, but *not at the same time.* If you are using WAP over GPRS on your class B phone and someone calls you, you have to decide whether you want to take the voice call and suspend the WAP session or just divert the call to voice mail. As we saw in the previous chapter, developers built this functionality into WAP, and WSP will suspend the session and resume it seamlessly once the voice call ends and the transceiver is free for GPRS. Now, consider that you are using a Pocket PC for your application, and your Pocket PC is GPRS class B-enabled (built in, compact flash, and so on). You might then use a Bluetooth headset in order to handle voice. For that device, you will likely have some wireless applications that are not WAP-based but rather were written by using C or Java. When this device receives an incoming call and the user accepts it, your application will experience an interruption. This interruption might be for just a few seconds up to hours (remember when you were a teen-ager?), and your application must be capable of handling this situation.

At first glance, interruptions such as these might be just a bit annoying, but they are not a serious problem for applications. Sure, there will be applications that crash, but the user will not mind restarting the application, right? The fact is that mobile users are much more sensitive to disruptions than desktop PC users, and the tolerance level will be much lower. We will discuss this topic in more detail in the next chapter, but now let's look at some examples that should make everyone concerned.

> Virginia Melovic is a professional musician who has an active lifestyle but still invests all of her money in the stock market herself. When not at practice, she spends lots of time researching stocks on her computer in order to find new bargains. In the limousine on her way to and from practice, she can receive alerts from her broker on her GPRS-enabled smartphone. One day, she is just about to arrive at the concert hall when her phone vibrates to let her know that something has come up. She pulls open the device and sees an alert from her broker that he has gotten some shares of a company that is going IPO tomorrow. The broker asks Virginia to use the application on the phone in order to confirm whether she wants any shares and how many she wants. Virginia becomes enthused and wants to purchase 2,000 shares immediately. As she types the number and presses the Confirm button, the limousine passes through a 20-second tunnel—during which the smartphone is not connected for half the time. As the connection rebounds at the other side of the tunnel, Virginia is anxious to know what happened. Did the confirmation message get through? Should she press the Confirm button again, or will that make the broker buy 4,000 shares?

In the future, this dilemma will be common for all *mobile-commerce* (m-commerce) applications as well. You'll buy something with your mobile device, but due to interruptions, you'll not know what happened.

- The server/receiving party might not have received the message or buy order.
- The other party might have received everything, but the acknowledgement may not have gone through to the user.

You might find it obvious that a transaction such as this one should be acknowledged, but sometimes the server-side software chooses not to confirm anything until all checks have been performed (credit card, membership, and so on). We recommend using the WAP model, where the request can be acknowledged even if the response has not been fully formulated. In other words, the user can first obtain an indication such as the following:

> "Buy order received; please wait while we check your credit card."
>
> "You have successfully bought 100 inflatable toasters. Thanks for shopping at Junks'R'Us! Click here to save this receipt on your device."

Those statements make the user much more comfortable while waiting for the final acknowledgment.

The next question is, "How should fatal interruptions be handled—interruptions that will not rebound?" For obvious reasons, the application will not know whether the connection will rebound or not, which makes the situation more complicated. The user might be heading from California to parts of the Nevada desert where there is no coverage, or the battery of the device might run out. For these kinds of interruptions, the application does not have much do until the user finally reconnects and resumes the application. For a WAP application, the server will suspend the session in this case and restart the microbrowser, and then the application will resume. At this point, the developer of the WAP application must make sure that the user first sees a deck that tells him or her the status of the account/order. The user should not have to spend a single minute trying to find out whether his or her transaction got through (this information should be more than obvious). The same general guidelines are valid even when you do not use WAP. The difference is that you will receive less for free and will have to implement the suspend—resume functionality yourself (or use the WAP stack from WSP downward, if the target device enables this feature). Generally, this process involves maintaining states in the client and server, which you can occasionally synchronize (for instance, when a crucial event happens, such as a purchase). A prerequisite for any of these actions is, of course, using reliable protocols (which means acknowledged WTP for WAP). When you use unacknowledged mode in WAP, you sometimes will encounter problems with the capability to recover from interruptions (even with short interruptions, such as a few seconds).

If your software runs on the device, you will have the advantage of storing the current state on the device. The client side of the application will know exactly what application has been presented to the user, such as whether a confirmation of a buy order has been shown or not. When coming back from an interruption, the application has a very good idea of where to start. In the example with Virginia, the client side of the application knows that she pressed the Confirm button right before the network connection was lost, and it can easily anticipate that she will want to know as soon as possible whether or not her information got through. A general rule for applications that are deployed on the device is therefore to store critical information about the application and its current state.

If the application is executing on the device, there are other means of handling interruptions as well. The client-side application, which needs input/interaction with the server, sends a request and awaits an answer. In modern operating systems, separate threads commonly implement these requests so that one request does not occupy the attention of the entire application (or even the entire operating system). In other words, you might use one main thread for the core of the application and use another thread to update the screen (and so on). In a wireless environment, it is important not to be dependent on the connection to the network, and those tasks that handle this procedure should use separate threads. As an example, in a chat application the writing of the messages should be totally separated from the part that manages the communication with other chat clients and servers. A user can keep typing even when he or she moves out of the coverage area for a while. Even if he or she presses the "Send chat message" button during this time of no connectivity, it should be capable of handling the send in the background (enabling the user to compose another message or perform another task).

Some devices and operating systems do not have concurrent threads at all (no multitasking, such as Palm OS) but overcome this deficiency by implementing a powerful mechanism by which the user can stop a transaction (request/response) without killing the entire application. If the user feels that he or she is in control, waiting for a delayed response is not too painful. Waiting becomes a hassle, however, when you do not know when and whether the application will recover and when you have no means to interfere. A successful application appreciates and utilizes the possibilities of having mobile, constant access to information but knows the limitations of this constant. If you know that the risk of a delayed response exists, will you put the entire application on hold while you are waiting? We discuss this topic more in the following section that explores perceived performance.

You cannot, however, only rely on the *operating system* (OS) to solve the problem for you. Instead, you must always think, "What happens if it takes 30 seconds to get a response or if I don't get a response at all?" Also, you must obtain a feeling for how threads or *Remote Procedure Calls* (RPCs) are handled in the platform that you have chosen. This hazard is not difficult to overcome, but if your application (and maybe also the OS) has been designed for Internet access where interruptions are very uncommon, you will have to be careful.

The effect of user behavior will be interesting to see here. If the user uses his or her favorite application while walking around and then suddenly gets interrupted, what action will the user take? If the task was really important, such as a purchase or a bank transaction, he or she might try to quickly get out of the radio shadow in order to finish the task. In some situations, such as when a user is in a moving car on the freeway, it will be just about impossible to affect the radio conditions (or maybe it will be the dawn of new problems related to mobile phone usage in cars that have panic breaks in order to stay in the coverage area).

In addition to the guidelines that we give here (which are specific for interruptions), you can achieve quite a bit by increasing robustness and perceived performance (as we will describe later in this chapter). Both of those remedies also help when you are dealing with high and varying latency.

Dealing with Latency

As new mobile systems emerge, one of the first questi[illegible]s, "How much is the maximum bit rate?" This question probably dom[illegible]ecause bit rate is one measurement of performance that is easy to unde[illegible]nd. On the other hand, when you are buying a new car, do you typically ask how fast it is—although it is easy to comprehend the meaning of speed? Although the maximum bit rates of wireless networks are steadily increasing and people always will want more bandwidth, many applications will not be limited by the available bandwidth. Especially for applications that have a high level of interactivity, the delay before replying to a request will be much more important. The delay between someone sending a message and the other party receiving it is often called *latency* and will be one of the most important parameters to consider when developing wireless applications. Throughout this chapter, we will use the words latency and *Round-Trip Time* (RTT) synonymously because they indicate the same thing, namely the delay (while RTT is the time that a message takes to get to the receiver and back) before getting a response to a request.

On the fixed Internet, RTTs are typically in the range of 10s of milliseconds (or a few hundred milliseconds for modem users and links that travel large distances). A common way of measuring RTT is to use the Unix command *ping*. This command also works from the command line in a *Disk Operating System* (DOS) window under Windows 95/98/2000:

```
C:\ping www.ericsson.com

Pinging 192.168.14.14 with 32 bytes of data:

Reply from 192.168.14.14: bytes = 32 time = 34ms TTL = 128
Reply from 192.168.14.14: bytes = 32 time = 32ms TTL = 128
Reply from 192.168.14.14: bytes = 32 time = 40ms TTL = 128
Reply from 192.168.14.14: bytes = 32 time = 25ms TTL = 128

Ping statistics for 192.168.14.14:
     Packets: Sent = 4, Received = 4, Lost = 0 (0% Loss)
Approximate rount trip times in milli-seconds:
     Minimum = 25ms, Maximum = 40ms, Average = 33ms
```

These results show the ping times from a computer in the MAI Stockholm labs to a computer that serves Ericsson's Web pages. Ping just sends an *Internet Control Message Protocol* (ICMP) message to the destination server and awaits the results. This particular message does not require any processing at the destination server, and the time that it takes to go back and forth should reflect the total time that it takes to travel that distance. As we saw in the example, the average time is about 33 milliseconds, which is fast. The slowest link to this server is probably in the range of a T1 line (1.5Mbps), and the links that are involved are very low-latency links.

To get a feeling for the latencies that your wireless applications will experience, we now perform the same command over an emulated GPRS environment. We perform this task by using the Global Application Test Environment (GATE) emulator that we will describe in more detail in Chapter 14, "Testing the Wireless Applications." The laptop that runs Windows 98 now talks to the Internet via the Linux computer that emulates the network. In a real situation, this laptop would have a GPRS PCMCIAcard or a Bluetooth connection to a GPRS phone. In this example, though, the GPRS radio is emulated inside the GATE. With an emulated 4+1 TS mobile, five other low-traffic GPRS users, no voice users on the same transceiver (TRX), and pretty good radio conditions (C/I = 18dB), we get the following result when we ping the location:

```
C:\ping www.ericsson.com

Pinging 192.168.14.14 with 32 bytes of data:

Reply from 192.168.14.14: bytes = 32 time = 450ms TTL = 128
Reply from 192.168.14.14: bytes = 32 time = 520ms TTL = 128
Reply from 192.168.14.14: bytes = 32 time = 360ms TTL = 128
Reply from 192.168.14.14: bytes = 32 time = 370ms TTL = 128

Ping statistics for 192.168.14.14:
     Packets: Sent = 4, Received = 4, Lost = 0 (0% Loss)
Approximate rount trip times in milli-seconds:
     Minimum = 360ms, Maximum = 520ms, Average = 425ms
```

As we can see, the average RTT is now 425 milliseconds, which is a lot higher than the fixed result. Now, let's see what happens if we bump up the number of GPRS users to 40 and the number of voice users to four. Now, you have 40 other users who are competing for the four time slots (four time slots occupied by voice users). The new result is as follows:

```
C:\ping www.ericsson.com

Pinging 192.168.14.14 with 32 bytes of data:

Reply from 192.168.14.14: bytes = 32 time = 900ms TTL = 128
Reply from 192.168.14.14: bytes = 32 time = 750ms TTL = 128
Reply from 192.168.14.14: bytes = 32 time = 602ms TTL = 128
Reply from 192.168.14.14: bytes = 32 time = 834ms TTL = 128
```

```
Ping statistics for 192.168.14.14:
     Packets: Sent = 4, Received = 4, Lost = 0 (0% Loss)
Approximate rount trip times in milli-seconds:
Minimum = 602ms, Maximum = 900ms, Average = 772ms
```

Figure 8.2 shows how the bandwidth usage increases dramatically as more users join the cell. This situation occurs because GPRS basically enables concurrent users to take turns at sending, and the more users who join the cell, the longer the latency for each user. We will discuss GATE, the tool that we use here, in more detail in Chapter 14, "Testing the Wireless Applications."

After seeing this example, we see more clearly that the latency is not only in the air interface propagation, but also in the processing of the traffic as the load increases. The processing in the network nodes takes time, and so do the retransmissions over the air. In cases of a high load, GPRS latency can be as high as several seconds (which can be devastating for chatty applications). 3G networks are expected to have lower latency, but LANs and radio networks are still very different.

Chatty applications are those that have a tendency of running off to the server all of the time, just to have the capability of enabling the user to perform the smallest task. Many applications do not even need the user to trigger requests

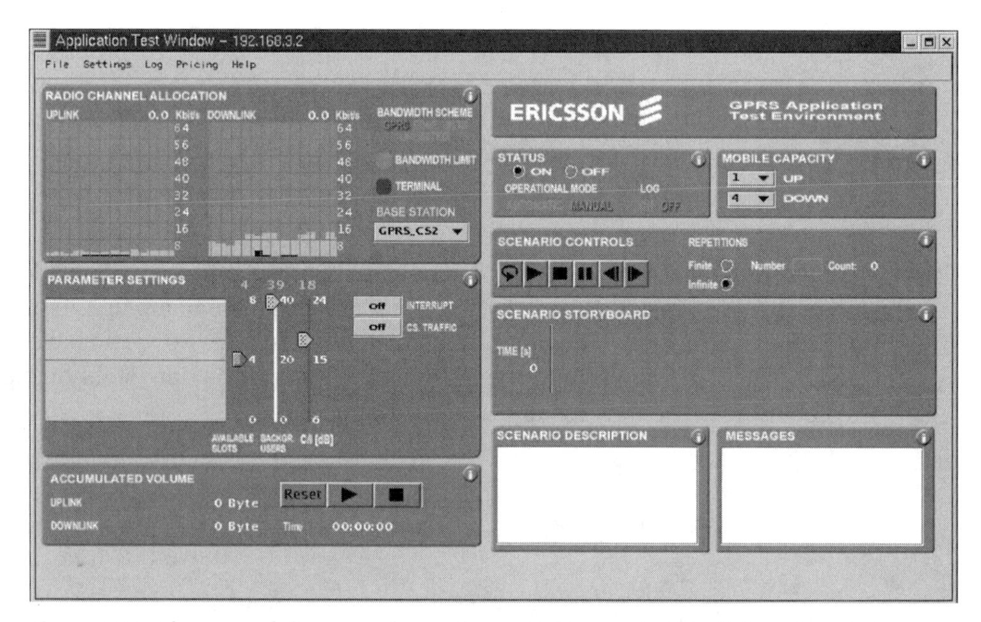

Figure 8.2 The upper-left corner shows the available bandwidth (light colored) in a loaded cell.

to the server; rather, they constantly check network nodes for new information and updates. One of the key reasons why WTP is transaction oriented instead of HTTP/TCP packet oriented is to minimize the chattiness (HTTP is a very chatty protocol). In this way, you can fetch one message (an entire deck) by using one request and response. This method is the desired way to write wireless applications, where you always try to minimize the number of interactions with the server and try not to start too many sessions.

Other means of combating latencies are the same as for interruptions: using separate threads for communications and keeping vital information about the client. All of these measures result in more robust applications.

The Effects of Packet Loss

In the previous chapter, we saw a significant increase in latency when the number of users increased. This situation shows that latency is not only in the air interface propagation, but also in the processing of the traffic. The air interface delays are mostly due to retransmissions that occur when data becomes corrupt or lost because of low signal quality. Lost packets on fixed networks mostly indicate that there is something wrong with the link. Most users never experience packet loss, however, because their applications use TCP (which covers the retransmission). In that case, the lost packet is retransmitted, and the user only feels a delay. For real-time applications such as streaming media and online games, we commonly use UDP instead, and packet loss becomes an issue. Even then, packet loss is generally equal to zero or close to it. The only time that packets become lost is when something goes wrong, such as when a router starts to toss packets because of congestion.

As we saw in Chapter 6, "Unwiring the Internet," we do not want TCP to handle retransmissions that are caused by errors over the air, and even some UDP applications might want rapid retransmissions over the air link (the user might not be able to notice the delay that caused this problem). If over-the-air retransmissions are turned on (in GPRS, turning on RLC-acknowledged mode), packet loss over the air is more likely to be visible to the application in terms of higher latency than loss of data as the data eventually arrives. The operator usually sets this parameter, which is likely to be turned on in most networks. If you have tried an interactive application such as Quake, when packet loss goes up by just a few percentage points, you know that this situation is something that you really want to avoid. Dealing with latency is much preferred.

MPEG-4 streaming is an example of an application that is likely to use UDP but that is very sensitive to packet loss. In MPEG-4, different packets hold different

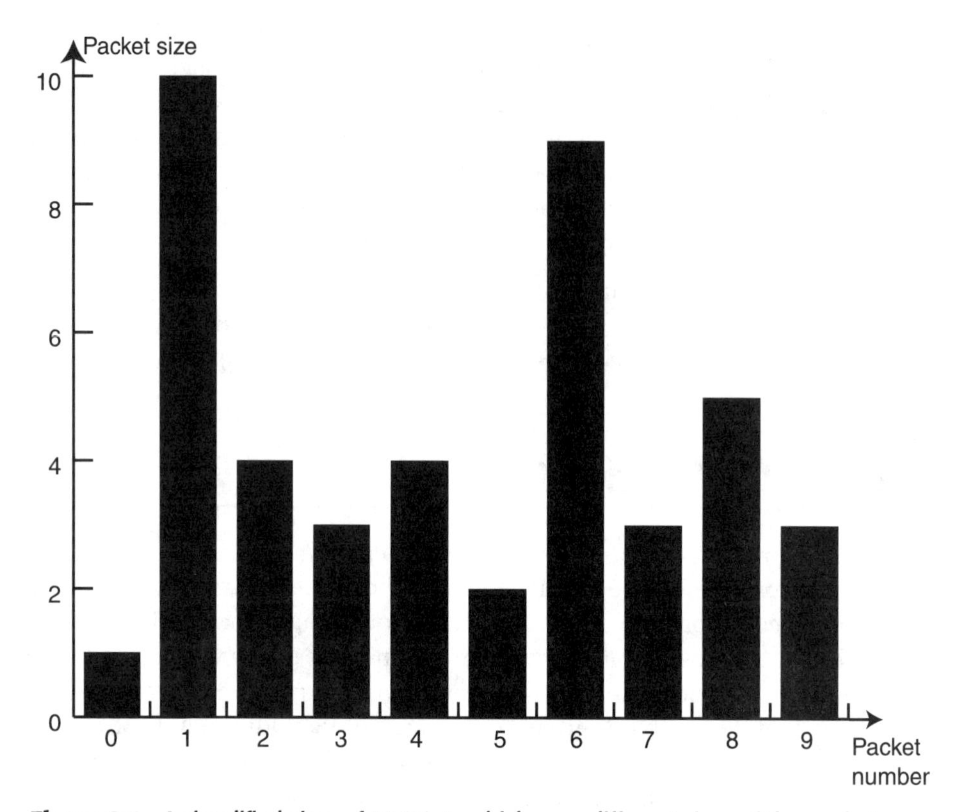

Figure 8.3 A simplified view of MPEG-4, which uses different sizes of the packets.

amounts of information, and there is the potential to prioritize important packets. Figure 8.3 shows the different amounts of information in different packets.

This figure shows how packets 1 and 6 hold major updates of the screen (for instance, at a scene change) while the others only hold information about how the screen should be updated. In other words, you do not have to send a static background; instead, you can send changes in the moving parts of the picture. While this example shows how you can optimize bandwidth usage, this example also shows some difficulties with handling interruptions and packet loss. If one or two of the smaller packets are lost, the user might not notice much of it. If, on the other hand, one of the major screen updates becomes corrupt or lost, the quality will be significantly lower (the screen will freeze or blank for a couple of seconds). A trick that you can use in the example of streaming is to delay the feed for a few seconds and always buffer some data ahead of time. This method has proven very effective in combating both packet loss and short interruptions. If the feed is not live, the streaming application can also utilize periods of higher bit rates in order to buffer ahead of time. With the irregular data flow shown in Figure 8.3, it would be catastrophic to lose the big packets. For example, you can imagine that this situation would cause a major disruption in the video if the second packet becomes lost (due to the mass of information that it holds).

For the majority of applications, however, where retransmissions cover losses, it is likely that packet loss will appear to most wireless applications as delays in the delivery of the packets (rather than as a loss).

Improving Robustness

Although issues such as interruptions, latency, and delay can seem difficult to handle, you can often address these problems by just being aware of them and adjusting your applications accordingly. This situation often leads to applications that show a general robustness and resilience against all sorts of problems, rather than individual optimizations for specific problems. As an example, let's consider an application that approaches the server in order to fetch some data but that enables the user to carry on with other tasks in the meantime (multithreading). This multithreading not only makes the application robust against interruptions (for example, when you are out of the coverage area for a while) but also makes it robust against other potential problems. A sudden increase in latency would, in some cases, not even be noticeable to the user. In addition, the same application will be appealing to users of the fixed Internet as well, because it will not suffer badly if the server is heavily loaded (an increase in processing time that produces longer response times) or even down for a brief period. In this example, just using separate threads for communications tasks greatly increases robustness (and most importantly, user experience).

We have examined multithreading and its benefits to overall robustness, but there are also other methods, of course. When we talk about improving robustness, we must take one step back and ask what we can do if *something* happens, rather than "what if *this* happens." Instead of finding ways to deal with interruptions specifically (although we often need to do that as well), you should look at how to make the application tolerant to any disruption, whether it is a long delay, packet-loss interruption, or something else. A good practice is to think things through even in the earliest phases of the application design and try to locate which modules/blocks of the application will depend on the network to different degrees.

The main difference with ensuring robustness in an environment (mostly in the operating system) that does not support multitasking is that you implement the remedies more in the details as opposed to during the early high-level design, as we described previously. Palm OS (3.5 is the latest release at this writing) is a good example of an operating system that does not support multithreading but that still enables you to create robust applications. The difference is the actual implementation of the code, where at all times you have to expect the unexpected. Send a query to the server, but keep generous timeouts and let the user make the choice of when to stop. Keep statistics regarding the time that it takes to get replies from the server, and adjust the communications properties accordingly (in other words, by being less keen on refreshing the cache or looking for updates to the server side). Many of the things that you can do in order to ensure that an application feels robust in a nonmultithreading environment involve that single word, *feels*. We will look more at how to improve this kind of perceived performance later in this chapter.

Improving Efficiency

Everyone wants to create efficient and lean applications for mobile devices because it is obvious that efficiency is crucial when developing for small devices and/or for low-bit rate networks. The first question that we need to ask ourselves, however, is what we mean by *efficiency*. Although an application certainly needs to utilize the memory efficiently and enable the CPU to rest as much as possible in order to save battery power, such improvements are out of the scope of this book. In this book, we concentrate on optimizing the wireless properties of the application; thus, the following parameters are of interest to us:

- Minimizing the amount of transactions
- Minimizing the size of the data that is sent over the air
- Minimizing the frequency of wireless network usage

We should note, however, that as this book title hints, the discussions involve applications where users are always online (GPRS and onward). Before, we saw two main groups get started with developing applications for GPRS and 3G, existing 2G mobile Internet developers, and software developers migrating for the PC platform and/or the fixed Internet. There are two main reasons why these efforts sometimes end up making their applications inefficient.

First, to a large extent, 2G mobile Internet means circuit-switched data where users will not pay more if 30KB are sent instead of 20KB by using 9.6Kbps (where the throughput is not limiting and more data could be sent within the same time). For such circuit-switched networks, the user is commonly billed for the time that the application is used, rather than for the data that is sent and received. In packet data networks, on the other hand, where users are more likely to pay according to the amount of data that is exchanged, things will be very different. Efficiency is the key. If the application is nonoptimized so that 40 percent more data is sent than is needed, this situation will be apparent to the user (whose phone bill might rise correspondingly). As we saw in Chapter 3, "GPRS—Wireless Packet Data," the billing scheme for each user depends on the choice of the operator in question. For the sake of simplicity, however, we will assume here that bytes are money and that the more the user sends/receives, the more he or she will pay. Charging users for used bandwidth instead of the time used demands a change in thinking, where you should carefully consider each transaction.

Second, users who are used to LANs and the fixed Internet are not at all used to thinking about how much data is sent, because the available bandwidth is more or less free as long as it is kept below the capacity maximum. The important thing has been to make the application work and get it to market as quickly as possible, not thinking about how a transaction of 250KB could be optimized to 150KB. How difficult this transition is depends largely on the product being developed. Is an existing application for the fixed Internet being changed to fit the mobile Internet, or is it a completely new application? Obviously, it is easier to change your thinking in order to accommodate new thinking if you can start with a blank piece of paper. For the optimization of existing software, you might find it especially interesting to look at the middleware solution, which we will describe later in this chapter.

You can often trace a lack of efficiency to massive protocol overhead or just plain excessive data transmission. The latter is especially common when you are porting applications from the fixed world, while protocol overhead often relates to using protocols that are not optimized for wireless.

Overhead

As we saw in Chapter 6, "Unwiring the Internet," the use of some protocols such as HTTP and TCP not only lead to problems with reaching desired speeds for the transmission, but also the protocol headers that are involved can create overhead that the user will pay for in the end. While HTTP/1.1 enables several requests to be sent over one persistent TCP connection, we do not always use this feature when writing the communications parts of applications. Do you know how to make this choice in your favorite programming language? Most of us just send a **GET** request to the server without caring how the data is sent to the client. Web browsers have the built-in feature of multiple HTTP requests so that you can fetch one page by using one TCP session (although it contains several objects). For other applications, it might not be as obvious to decide which HTTP sessions should use the same TCP session. The amount of data that is being fetched might be different each time, and at this writing, it is hard for us to estimate how we will use the code.

One example is when a client wants to synchronize data, such as sales information or a calendar, with the server. Typically, this kind of synchronization involves making sure that a large number of records (address entries, database fields, and so on) are updated. The intuitive way of performing this task is to use HTTP (over TCP or some other protocol) in order to send each record to the server. In other words, every record—even if it is just 20 bytes—receives its own HTTP command with the associated overhead (if you use HTTP/1.0, you must start a new TCP session, as well). The overhead for each such HTTP transmission varies with how much data each record holds, but this amount could be many bytes. The pain is not over, because the overhead will get even bigger if the bandwidth is really low (as we described in the section about low bandwidth issues). This approach of application design can lead to hundreds of percentage points of overhead, which effectively means that the user is paying several dollars per each dollar's worth of useful data. Most users will not be happy if they realize this fact.

We can dramatically improve this example by assembling several records into each transaction. Setting a parameter that determines how many records are assembled into each request can enable this function, and then you can set the parameter dynamically by the application or statically through initial configuration. The final phases of testing the application, where you use an emulated network, is an excellent time to fine-tune such settings. Another way is to create an additional buffer after the final transmission buffer of TCP. You can then set this buffer to the desired packet size so that records are collected before they are all sent together in the same packet. AvantGo, a company that has extensive experience with optimizing applications for Cellular Digital Packet Data (CDPD) and GPRS, introduced this method some time ago. Its final phase of CDPD testing was then used to fine-tune the optimal packet size. Many developers ask me what the optimal packet size is for wireless networks, but providing an answer is almost impossible. Generally, you should go for larger packets if you have a lot to send (such as synchronizations, downloads, and so on) and use smaller packets for interactive applications (chat, games, and so on). The smaller the packets, the bigger the impact of latency—so the best way is to make this parameter adjustable and not decide how to configure it until the last phases of testing.

How much the overhead is for each of these protocols depends not only on how big the data payload is (a small payload per packet means big overhead) but also on the radio conditions with the associated retransmissions. In an example of fetching a Web page (by using HTTP/TCP) from a server over an emulated GPRS network, the accumulated traffic in uplink and downlink can increase by 50 percent for a small page as the available bit rate decreases from 20Kbps to 5Kbps.

This situation can be harder to circumvent, however, because the increased traffic is mostly retransmitted packets and acknowledgements. Setting generous timeouts for TCP (if that setting is available) helps a bit because it leaves more of the retransmission responsibility to protocols such as RLC that can cover retransmissions over the air link. As always, this situation depends on whether the platform in question enables you to affect this parameter. This dilemma also teaches one of the more important lessons of optimizing applications for wireless: Gains in persistence often mean a loss of efficiency. For low bit rates, we can force the data to go through by repeating the retransmissions of lost packets, and we *will* get more data through—but at the same time, we will cause more overhead. This kind of persistency also affects the battery life of the device, because it has to be more active. Again, this situation is a tough trade-off to make, and the final tuning of the affected parameters should not be made until the later phases of testing, where you can investigate how the performance will be for various network conditions.

Excessive Data Transmission

While protocol overhead is a difficult issue that often results in difficult decisions and trade-offs, excessive data transmission is more about common sense. Here, we use the term to describe applications that just send lots of data back and forth between the client and the server, causing the user to wait longer and pay more. This situation could be due to not storing enough critical information on the client and having to continually fetch it from the server.

An example could be a Java game where users race against each other over mobile networks by using client software that is installed on the devices. While you might perform the initial installation by using a WAP interface, where the user clicks a link in order to download the program, you do not have to download the program again before every game (provided that the devices can have software persistently installed on them). A better solution might be to send one request to the server at the start of the game, letting the server know what version of the software the client is using. The server can then complement the game with updates (if necessary), and in that way, it limits the amount of data that is sent. Not only does this process save time for the user, but it also saves money. Although some 3G networks might look temptingly fast (and downloading a 500KB application each time might sound like an easy task), the users who have to pay for those bytes might object.

Reducing the amount of data is not as hard as many other aspects of optimizing for wireless networks. Developers, who are educated in this area, will find that being aware of this problem should be enough to deal with it efficiently.

Using Compression

In the drive toward minimizing the amount of bits and bytes that we send over the air, you will obviously find it interesting to examine how compression can help. Compression is already widespread and proven for fixed-Internet applications, and you can apply many of those techniques to wireless networks. Networks themselves, however, will use some compression technologies (such as v42bis), leaving the applications developer at the mercy of wireless infrastructure vendors. The same is also true for the header compression of TCP and IP headers—a process that often takes place in the operating system. Others, such as .jpeg, .zip, and so on give the developer the option to add content-specific compression, which can significantly enhance performance.

When you are talking about compression for wireless, v.42bis should often be the first topic on the agenda. The ITU-T proposed the v.42bis compression standard (at the time, the ITU-T was called the International Consultative Committee on Telephony and Telegraphy, or CCITT) as an addition to the v.42 error-correction protocol for modems. The thought is to increase bit rates over wireless networks by using a general-purpose compression method that does not depend on the content. The compression algorithm continually monitors the data to be sent and checks it for compressibility (it can also choose to send the data uncompressed or compressed). If the data is already compressed, v.42bis will not likely be capable of compressing it more. One problem with v.42bis is that some patents cover it, and compression newsgroups on the Internet complain about the complicated licensing terms. V.42bis is commonly implemented in modem hardware and thus would be a part of the mobile telephones and base stations. You can also implement v.42bis in software, however. You can find the specifications at ftp://ftp.fdn.org/pub/Library/Ccitt-standards/ccitt/.

V.42bis support is not mandatory in GPRS release '97 implementations, and at this writing, it is unclear when we can expect widespread support. You can activate compression by setting the <d_comp>-parameter in the PDP context, for instance, by using the +CGDCONT AT command (as specified in GSM 07.07). Chapter 3, "GPRS—Wireless Packet Data," explains in more detail how to use AT commands.

In addition to this kind of general-purpose compression, numerous solutions are available that are specific to the media that you are using, such as .jpeg, .mpeg, .gif, and so on. Many of these solutions introduce a loss of information in the process, however, assuming that the user will not notice a dramatic difference. Whatever method you use, you should look into the power consumption that is introduced and the CPU power that is necessary. .mpeg, for instance, was originally intended for devices that have dedicated encoding/decoding hardware and puts a lot of strain on a mobile device. This concern might not be an issue in the future as devices with built-in multimedia hardware emerge, but this issue can be important for developers.

Sometimes you can implement simpler methods, rather than going as far as introducing an entire compression engine. The kind of encoding that WSP uses, where its well-known headers each are represented by a code, is an inspiring method that you can use in other cases as well. A chat application, for instance, might have a number of control commands and special characters that appear frequently. Instead of sending those over the air, an encoding into some negotiated codes could replace it and increase efficiency.

Using Caching

While there are many sophisticated ways to improve performance over wireless networks, one of the simplest ways is also one of the most efficient ones: caching. Most users today are familiar with the concept of caching, because they know that the Web browser not always gets the information but that the Reload button will force it to perform this task. The most famous analogy involves going shopping. You store the most commonly used groceries in the refrigerator (cache) so that you do not to have to run down to the store every time you make dinner. For wireless applications, caching is utterly important—and the performance gains are often impressive. WAP uses caching extensively and also enables the application developer to control how individual objects are cached (although this task is not always easy) by setting expiration times (and so on). The size of the cache varies between different WAP devices and is likely to continue as such. For WAP developers, it is essential to test the actual interaction of the application and the cache. Objects that do not need frequent updates *should* be cached, and only testing will show whether things are displayed as they should be.

You can also implement caching on the client side manually in the application software, but if it is not a browser application, the difference between a cache and a smart management of client data is infinitesimal. Some middleware solutions include a general-purpose cache that stores objects that you frequently request. The gain of such a solution varies greatly with the application that you use, but this method is an easy way of gaining performance.

Although client-side caching is the obvious way to enhance performance, server-side caching will also become more and more pervasive. With server-side caching, the main gains are not as obvious because it will not transport any less data over the air. The gain is instead load distribution and making life easier for the application servers. Some applications are peaky by nature, meaning that the usage is extremely high at a few peak times and low most of the time. This situation puts the operator or service provider in an awkward position, forcing him or her to decide whether to deploy tons of application servers that can handle peak traffic (remaining unused most of the time) or just to have a smaller amount of servers that are incapable of handling all of the traffic at a peak time. An example is the typical traffic information service that gives users instant mobile access to the latest traffic and accident information. This application will have millions of hits per hour in larger cities during commute hours, while not many will use this service in the middle of the day. Using server-side cache proxies at strategic positions will enable the clients to receive the most common requests from the cache proxy instead of the application server. As a result, the operator does not have to invest heavily in excess capacity just to handle peak traffic. An additional benefit is that the cache proxy is likely to be capable of serving each user more quickly because it is closer to the users.

You could place server-side caches/proxies in many different places (for instance, on the service network). We describe applications architectures in general, including the service network, in the next chapter.

Buffer Issues

One of the mandatory methods for coping with irregular flows is to provide buffers that are big enough. During bursts of large amounts of data, the buffer can smooth this process by storing the data until the destination can process it. Especially with the advent of high-speed 3G networks, where throughputs are in the range of hundreds of kilobits per second, it becomes important to have buffers that can handle these bursts. The problem is even more general than that, however, and is often caused by a destination device whose CPU is incapable of handling the high bit rates. The processing of the data can sometimes be very power consuming.

The problem occurs when the bandwidth (as well as the latency) is high. You can measure this level by multiplying the bandwidth (bits per second) by the RTT (in seconds). The result is the capacity of the round-trip route between the sender and the destination. The bandwidth-delay product is measured in bits and is used as a measurement of how much data the connection maintains in the loop at one time.

On a 3G network that has lots of packet loss, the RLC layer will retransmit the lost packets (which, as we described in Chapter 6, "Unwiring the Internet," provide an additional delay in the transmission). As a result, the bandwidth will be in the range of hundreds of kilobits per second, and the RTT will be a couple of hundreds of milliseconds. As an example, let's say that the RTT is 400ms and the bandwidth is 400Kbps. The bandwidth-delay product will then be 400,000 bits per second $\times$ 0.4 s = 160,000 bits = 20KB. Now, let's assume that we are receiving the data on a 3G-enabled PDA. We will need to have more than a 20KB buffer—preferably even much more in order to handle the retransmissions. This amount is quite a lot, considering that some desktop operating systems have less than half that value as a default buffer size for incoming packets (see Figure 8.4). Desktop systems were built for high bandwidth but not at the same time as high delay. Therefore, the high-speed wireless systems will set new requirements, and we will need more research in this area.

Devices that have built-in 3G functionalities are likely to cope with this issue well because they were built with these scenarios in mind. The big question is how legacy PDAs with Bluetooth snap-ons will manage. Developers should consult the support resources of the targeted device in order to obtain more information.

Improving Perceived Performance

The fact that many companies are technology driven is nothing new, and people who know the technologies well are the ones who make the decisions. It is

Figure 8.4 Big pipe, small receiver.

also a well-known fact that these decisions are not always in the best interest of the end user (instead, they often only benefit the company itself). Examples such as the PAL TV system and the VHS video system show that the best technology seldom wins. Although we will not go into too much detail about the business aspects in this book (apart from the last chapter), we will say that a developer must never forget about the user's opinions. How optimized the transmission is or how high compression rates are does not matter if the user's experience is horrible. You can achieve a good user experience in roughly two ways: improving the real performance (compression, robustness, and so on) and improving the perceived performance. Perceived performance here means the measures that you can take in order to make the user feel like the application performs well (the things that might not be possible to prove in figures but that still contribute to the overall performance).

Keep the User in Control

The most important feature of perceived performance is to keep the user in control. There should never be a situation where the user sits and waits for something to happen without any possibility to affect things. Users should be spared from the classic hourglass wait for a task to finish, where pressing buttons in panic will not matter at all. This feature is especially important for those operating systems that do not support multithreading, so that the one task has to finish executing before you can do anything else. If the user does not have a chance to abort a task at that point, then the entire device will be locked until it finishes. A simple Abort button can be the difference between good usability and a useless application.

The developer always has to strike a balance between making the application persistent while keeping efficiency, as we mentioned previously. For some applications, it is appropriate to let the user influence this process. If the connection quality is really low and it is hard to get data through, only the user knows whether he or she wants to pay the price (in terms of battery life and generated traffic) of making the connection mechanism more aggressive. Figure 8.5 shows an example dialog that gives the user the option to control persistence.

The difficult part is, of course, how to communicate that the user might pay more if he or she tries harder. Some operators might offer a flat-rate pricing model, and then the cost of massive retransmissions will not be higher for the user. Even if the middle button in Figure 8.5 is skipped, the user is still in control and the usability of the application is higher. You will find it mostly wise, however, not to rely on dialog boxes or any events for Abort or other control buttons. A better way is to have a generic way to stop any task that does not rely on where in the execution the user is. It is always hard to anticipate correctly when

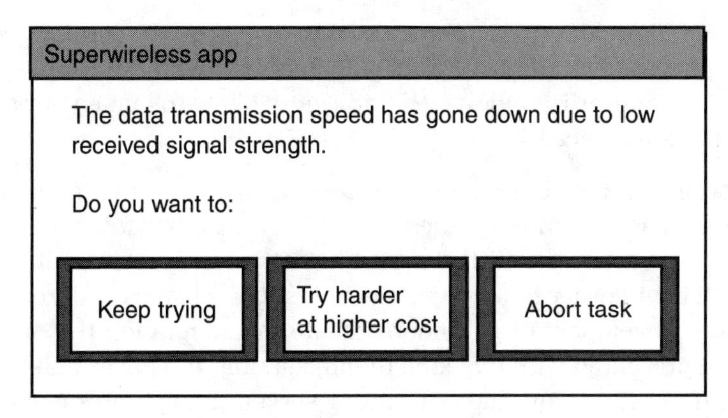
Superwireless app
The data transmission speed has gone down due to low received signal strength.
Do you want to:
Keep trying
Try harder at higher cost
Abort task

Figure 8.5 Keeping the user in control with extended dialog boxes.

someone will want to get out of a task, and the controls should be as generic as possible.

Interruptions will likely be the most common issue that requires applications to place the user in control. After all, in some situations the signals cannot reach the user at all, and in that case, there is not much to do but to make sure that the user still feels as comfortable as possible. If the application has a software client installed, it can be wise to let the user go into an offline mode as soon as an interruption is detected. The hard part is that the application rarely knows how long the interruption will continue; consequently, it does not know whether there is a point in going offline. Most of the time, the user has a much better view of this situation and can make better decisions. Take the example of a user who has a GPRS class B terminal where a data transmission can be interrupted by an incoming voice call. If the user is checking his or her mail when someone calls, it is of course preferred if he or she can keep the mailbox working (but in an offline mode as the phone call is underway).

Overall, it is recommended to have the possibility to switch between online and offline modes for all applications that have a client software part.

Keep the User Informed

A second remedy for the loss of a connection is to keep the user not only in control, but also informed. Compare watching your screen, which says "Fetching your valuable data, 15 seconds to go," or watching a lonely hourglass. If the users know that progress is being made (or if they receive some indication why there is a delay/halt), the user is much more likely to put up with the wait. Then, combining this information with the previously described control, the screen will say, "Fetching data, currently slow progress the last 20 seconds. Wait some more or click Retry or Cancel." This screen still does not present good news, but the user is empowered and is less likely to curse your application and reboot the device.

The information can also be as simple as a progress meter that tells the user how much data has been downloaded, how long a task has taken to execute, and so on. This meter can tell the user what other node on the network to which it is talking in order to let the user understand what procedure is taking the most time. Developers often underestimate the adaptation that users can develop when presented with this kind of information. If it always takes a long time to get information from the positioning server, the operator can be notified and improve the situation (will the operator like this kind of power in the hands of users however?). If the user downloads a form in an application, he or she can be presented with a progress meter that informs him or her about how much data has been downloaded and how long it has taken. We have seen how users can quickly learn about how long this kind of task should take and how much data needs to be transferred. The user can then make the decisions, such as how long to wait before deciding to give up.

A connection display that shows the user what service is activated and what servers the application is talking to helps empower users to make better decisions through information. The challenge is to put enough information and control in the hands of users without making them confused. You do not want dialog boxes with multiple technical choices to emerge during each part of the application execution. After all, most of this process comes down to common sense, and you can verify your actions through working with test groups of different technical knowledge levels.

Middleware Solutions

Many of these solutions require lots of knowledge and work on the part of the applications developer. With this many things to consider, it might not be appealing for everyone to go deep into details and combat all of the different aspects of developing for mobile networks. We already mentioned WAP as one way of avoiding many of these struggles, but there is another range of products (none of them directly competing with WAP) that help developers in a generic way: middleware. Finding a generally accepted definition of the word is difficult, but after looking at a number of sources, I can make the following suggestion:

Middleware is software (often denoted as a platform) that mediates between the network and the application and that enables seamless communication over heterogeneous networks.

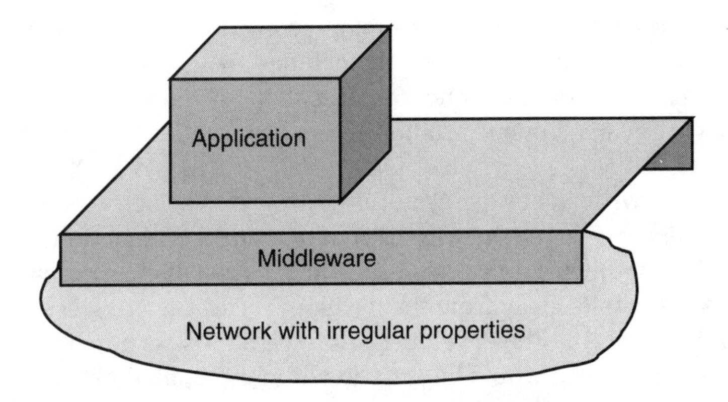

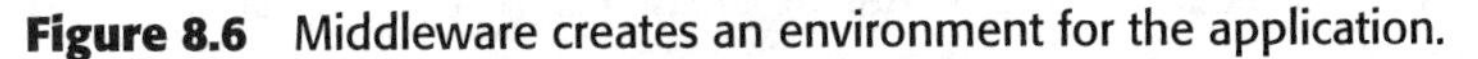
Figure 8.6 Middleware creates an environment for the application.

Figure 8.6 illustrates how the middleware lies on top of the network with all of its irregularities and creates a smooth environment for the application.

Middleware products generally aim at simplifying the use of networks and devices by supplying development kits with supporting client and/or server software. There are a number of different ways in which you can perform this task, and the needs of the developer determine the optimal choice. Today, there are many middleware solutions that have numerous features. From the application developer's perspective, I divided them into three groups: total middleware, detached middleware, and development studios.

Total middleware. Here, the design of applications is very much simplified and the developer can concentrate on supplying content. Some use a kind of drag-and-drop interface and others offer consulting services where pure content (databases and so on) is turned into a mobile application. The middleware then takes care of the adaptation to and optimization over different networks and even the presentation to the user. This solution is very good for pure content providers that just want things delivered as simply as possible. The drawback is the major lack of flexibility and differentiation between different users of the same middleware.

Detached middleware. For applications that were developed for HTML browsers and fixed Internet usage, a client-server middleware that performs caching, compression, and so on is suitable. Here, the middleware and the application are two separate software products, and you could run the application alone. This feature gives more flexibility and control over the user interface for the developer. Common features include data compression, header compression, and caching. A potential drawback is that the optimizations are very generic and are not tailored for the application. This situation generally gives less-efficient results.

Development studios. Most software developers are used to working with a development studio product that not only includes an advanced editor and debugger but that also contains other features that make programming easier. Some entire development studios are made specifically for wireless, and some existing ones have wireless features already included. As a result, the developer can write his or her C and Java programs as before, but he or she must use library calls for the wireless communications functions. The middleware then optimizes those requests and calls as much as possible and abstracts the details away from the developer. This feature gives lots of flexibility and control, but the developer has to do almost as much programming as he or she would have to do without the middleware. Another potential drawback is that both the client and the server usually need to have the middleware installed.

Of course, the question arises of who should use middleware. First, you need to determine whether it is possible to use middleware at all. If the desired middleware needs client-side software, the device of course needs to support that (open platform). Middleware can generally shorten the development time and cost at the expense of flexibility and differentiation. Today, with our shortage of skilled *Information Technology* (IT) personnel, it might be valuable not to have to hire skilled programmers but rather to just focus on the content and use middleware for the presentation. This situation especially applies to those companies that just want another channel for their content and/or to have legacy applications from the fixed Internet/LANs. For content developers who want to start from scratch and deliver content to mobile devices, however, WAP is likely the best solution.

One segment where middleware has proven successful is in vertical enterprise applications. This success is mostly due to the specialized devices that might include bar-code scanners and other add-ons. For those devices, it is a huge time saver to have a middleware solution that enables developers to quickly create advanced applications without having to worry how a bar-code scanner works. As the mobile Internet becomes pervasive, we are likely to see an increased market for such specialized devices and applications.

Overall, the price for using middleware is flexibility. Differentiating applications from the competition becomes more difficult, and there is a high dependence on the device properties. The choice is in the hands of the developer.

Summary

We can attack the problems in wireless environments in two ways: improving real performance and improving perceived performance. Real performance involves handling interruptions, latencies, and avoiding overhead. The developer should avoid going into too much detail; rather, he or she should concentrate on making the application robust and efficient. These remedies make the application more fit to handle the unexpected and to perform better overall. The perceived performance produces improvements that are hard to measure but that help the user have a better experience. This concept includes keeping the user in control and informed and keeping the user's needs in mind at all times. You can use middleware to enable a third-party product to overcome some of these difficulties. The application type as well as the device and operating system used are parameters to consider when you are deciding whether or not to use middleware.

PART THREE

Applications and Their Environments

CHAPTER 9

Application Architectures

When we described wireless networks in Part One, we examined most of the aspects that are relevant to the applications developer (except for where you can implement applications and how). One reason is because this area is so important, and we want to give it plenty of space. Another reason, however, is that we are migrating to a new way of implementing applications on wireless networks. Here, we use the term *application architectures* to denote the way in which applications connect to the mobile network and to other components.

Traditionally, technicians have implemented applications such as voice mail, *Short Message Service* (SMS), and so on as tightly connected to the nodes and switches of the wireless network, and these applications are specific to a certain mobile system (GSM, TDMA, and so on). This situation creates a very rigid architecture, where third-party applications developers have little or no chance of adding applications. In this chapter, we will examine how we can implement applications in *second-generation* (2G) and 2.5G systems and how we can accomplish this task on *third-generation* (3G) networks. In addition, we will take a closer look at the *Application Programming Interfaces* (APIs) that developers can use to access the features of mobile networks.

Architectures Now and in the Future

As we examine the mobile evolution and the introduction of new and improved systems, at first glance we tend to focus on the radio and network characteristics. Everyone wants to know how high the bit rates will be and what kinds of functionalities the handsets will have. Many other things have to come into place, however, in order to make the mobile Internet as flexible as possible. Remember that telecommuniations networks were traditionally built to carry voice and just a few vertical applications. Compared to the fixed Internet, these networks are secure and extremely reliable but also extremely rigid. Third parties have not found it easy to add functionalities and applications that are limited to SMS, voice mail, and other features that are tightly integrated into the existing networks. One system had a set of applications that was incompatible with those of other systems. Thus, a *Code-Division Multiple Access* (CDMA) user could not access an SMS center for a *Global System for Mobile communications* (GSM) network, and a fixed-line user had an answering machine that was different from the voice mail of his mobile telephone. Figure 9.1 illustrates this architecture of services that are dedicated to the individual networks.

When developers added data functionality (mostly circuit switched) to 2G networks, these limitations became more and more visible—especially with the rapid growth of the openly designed Internet. The Internet facilitated an enormous growth of applications and content, and it was extremely easy for anyone (user, content/applications developer, and so on) to connect to the Internet. With the advent of wireless packet data networks, the telecommunications

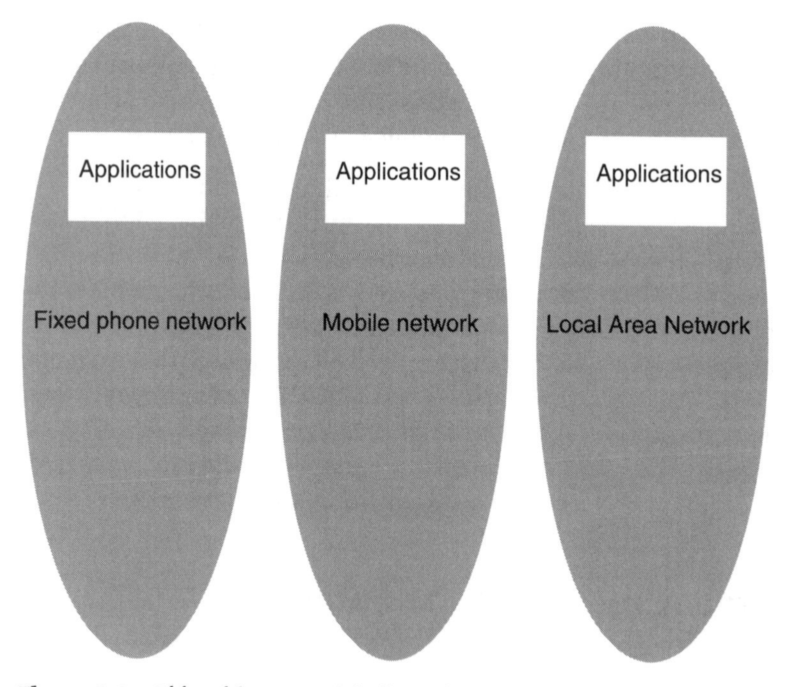

Figure 9.1 Old architecture of dedicated applications.

industry realized that the fixed-Internet developer community had to be mobilized in order for the mobile Internet to take off. A key issue that needed to be resolved was how to introduce an open and flexible application architecture to mobile networks without jeopardizing security and reliability. Another important aspect was to make applications for one network available to others as well. Someone might develop a banking application for *Wideband Code Division Multiple Access* (WCDMA) users, and it should then also be possible for CDMA2000 users to access it, as well.

The solution is a horizontally layered architecture that we touched upon in Chapter 4, "3G Wireless Systems." The three main planes are transport, control, and applications (services), as shown in Figure 9.2. We now want to dive deeper into the application layer and see how it consists of many components (more than just WAP gateways and application servers).

We will describe the three planes in detail:

Application plane. Not only are the applications located here, but a number of nodes that facilitate the new services also reside in this location. These applications include positioning servers, WAP gateways, and so on, as we will describe in the Service Network section of this chapter. As a result, mobile networks and clients or servers on the fixed Internet can all access these applications (provided that the owner of the service network allows

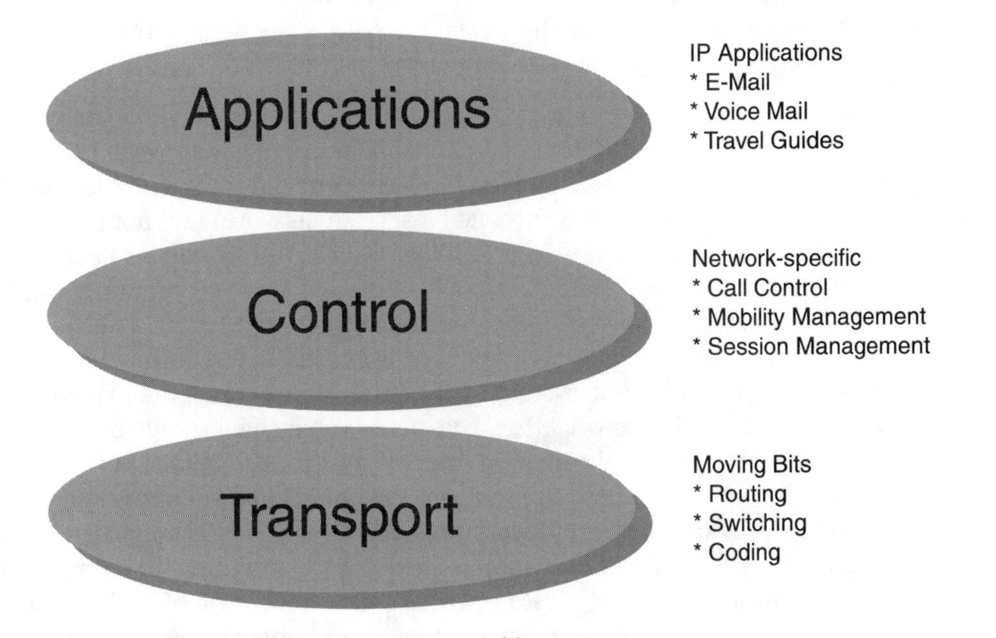

Figure 9.2 A new, layered applications architecture.

this access). The applications can be anything from browser-based WAP applications to any generic, IP-based application.

Control plane. This layer handles setting up calls, tracking mobiles, and managing billing information. Practically all of the intelligence of the mobile networks resides here.

Transport plane. Once the control plane decides to set up a call, the actual transportation of the bits takes place in the transport layer. This process includes all sorts of routing, coding, and switching. Some functionalities that previously took place on the middle of the mobile network, such as voice transcoding, can now instead be moved to the edge of the networks. This feature makes things more efficient and easier to control for the operator.

This new architecture provides the opportunity to divide the networks into logical entities based on functionality rather than on system technology (GSM, CDMA, and so on). All three layers are based upon open interfaces, which open the possibility for third parties to enter the game. If there were no open interfaces, the applications would have to terminate the telecommunications protocols and communicate with them directly. This process is not only inefficient and difficult, but it also directly limits the application to a particular network. The nodes of the application layer that offer these APIs now perform this dirty work, and the application developer gains access to high-level, network-independent functions. In the *Service Network* section of this chapter, we describe the service enablers that facilitate this process.

Note that these layers are logically separated, which means that they can be physically located at different places (not a requirement, however). Therefore, it would be possible for a company to start a business as a service provider, operating an application layer (called a service network) and not have to have any wireless infrastructure. This service provider offers services and applications, rather than the traditional mobile subscriptions that primarily offer network access. The service provider could then optionally choose to sell subscriptions that include network access as well and then buy this capacity from a traditional mobile system operator. In other words, the existing operators will have to decide what their role in the value chain should be. Should the operator not only offer the bit pipe (transporting traffic) but also be an *Internet Service Provider* (ISP) and offer services on top? Some operators will try to be one-stop providers, offering everything the user needs (including terminals, applications, and Internet and mobile system access). Other operators will take pride in owning global wireless networks and will focus on offering mobile system access. Today we can already see how some network operators divide their company internally in order to have one part working with the applications offerings and another part working on the actual mobile network. Later some of these players are likely to split their companies accordingly. The model that will be the most successful will depend mostly on its execution, and we will see winners in both camps. The important thing for operators is to actually face the change and set a firm strategy that outlines where they should be in the value chain. We revisit the business aspects of the mobile Internet in Chapter 15, "Getting It All Together."

Before we plunge deeper into this new architecture and the service network, let's look at how the applications are typically installed in 2G/2.5G networks.

Today's Application Architectures

If we use the new way of designing networks, we cannot design them overnight. Many GPRS networks will initially use the old way of integrating applications, putting servers deep within the mobile network. Although all operators have chosen different network designs, the basic outline is often similar. Mostly, there is an application/service LAN connected to the mobile network, as shown in Figure 9.3. We show a GSM system here, but IS-95 and *Time Division Multiple Access* (TDMA) systems are designed similarly.

In the figure, we see that the LAN where applications reside is tightly connected to the mobile system. The SMS-C, for instance, is connected to both the application LAN and the *Signaling System 7* (SS7) signaling network. In other words, there is a high barrier for third-party developers to develop applications for these systems. Not only that, but gaining access to system-specific features such as call control and positioning is close to impossible. (There are no APIs available.) In addition, this network design is not suitable for large subscriber

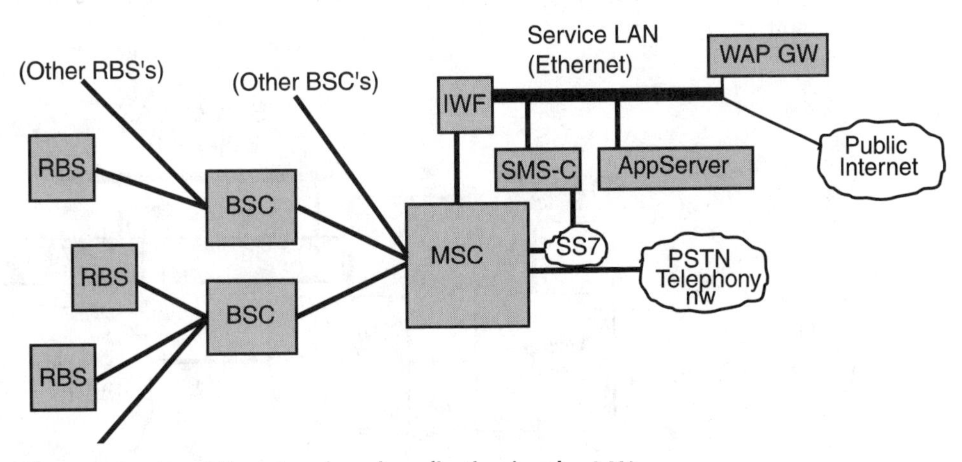

Figure 9.3 A mobile network and application/service LAN.

bases of applications. An operator that starts with tens of thousands of applications subscribers and suddenly receives millions will run into difficulties. The system scalability will be poor, and adding twice the amount of applications servers will only render a tiny percent increase in capacity. Finally, there is the issue of software upgrades to the components of the service LAN. Will adding a new application mean an interruption of service to other applications on the same application server (or even applications on other servers)? There is clearly a strong need to build the architecture for mobile applications, from the bottom up, in order to handle these challenges.

We expect GPRS to be a transition technology for application architectures in the sense that some operators will start with the old architecture and later migrate to full-blown service networks. Figure 9.4 shows a GPRS system that uses the standard 2G applications solution.

As illustrated in the figure, the GPRS network is closer to the desired IP-based application architecture, with the GGSN (which uses IP) directly connected to both the Internet/ISP and the service LAN without an Inter Working Function (IWF). This approach does not solve the scalability and robustness issues either, however, and the need for a dedicated application architecture (the service network) remains.

Introducing the Service Network

As the mobile Internet becomes pervasive, it also becomes more and more obvious that it is not an isolated network. On the contrary, there will be great synergies in doing some task of an application with a desktop PC (and some with the mobile device). You can book a complicated trip through a travel agency or directly on the Internet by using a desktop PC. At least, the first time that you

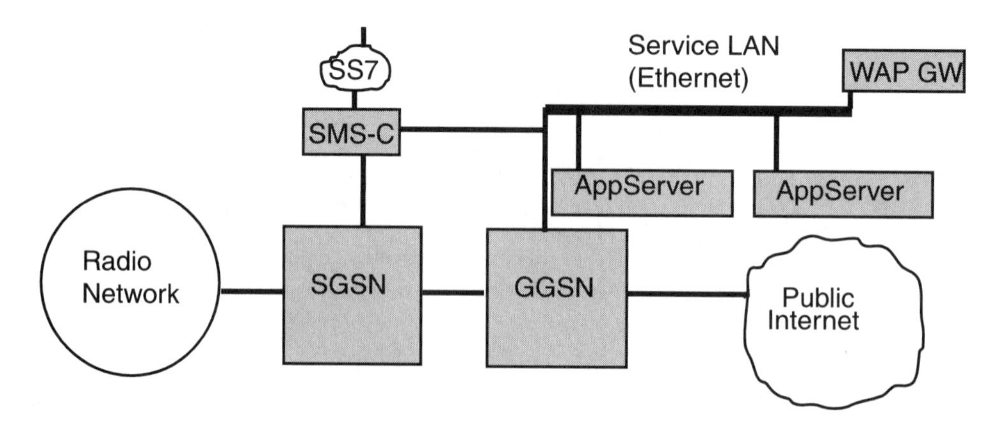

Figure 9.4 A GPRS network and application LAN.

perform this action, it requires lots of text entry because the payment and mail information need to be entered. Therefore, a nice 21-inch color monitor with a large keyboard is preferred over a 10-by-2 cm telephone display. Once the itinerary is set and the user is on the move, the lightweight phone is an excellent tool for checking confirmation numbers, hotel addresses, and so on.

As we saw in the previous example, accessibility from a variety of networks (fixed or wireless) is essential to this new architecture. Some of the other drivers are as follows:

Scalability and robustness. The network design should be suitable for any number of users, and it should be convenient to small with just a few nodes and add more nodes as more users join.

Flexibility. The capabilities of the network should be easy to expand in order to include new value-added services as they become available. For instance, one operator might decide to start without location-based services and add support for that later. An application server might initially contain only one application, but later on you might need to install more. This situation should in no way impact the current service to subscribers.

Security. As the applications become logically separated from the mobile network, you must ensure the security, integrity, and privacy of users and their data. The closed operator environment of 2G systems was what made security easier, and a new open architecture means some new security challenges.

Low cost of ownership. Despite high demands on performance, the service networks need to be easy to maintain and evolve in order to ensure widespread usage. There will be many new players in the field of service/application operators, and their barrier of entry should be low.

Personalization and service roaming. Wherever a user travels and whatever network he or she is connected to, the offered services should appear consistently. This statement also means that an application that knows your preferences in France also knows your preferences in Dubai. We commonly call this vision the *Virtual Home Environment* (VHE), indicating that any network should feel just like home and that your favorite services should be available no matter where you are.

Wherever possible, the architecture is built on existing technologies and standards. 3GPP is developing additional standards and will support these when they become available.

Architecture Overview

Below we will show one way of building a service network. It should be noted that individual solutions might differ, but the general thoughts and interfaces are the same. The service network is based on IP, which means that you can implement the connecting infrastructure by using standard datacom components. Everything is based on components in order to obtain a modular architecture, where it is up to the application what parts to use. Figure 9.5 shows an example outline of the architecture with its main components.

This picture might look like a giant leap from the GPRS architecture that we described in the previous subchapter, but it is rather a more structured way of designing things. The core network at the bottom of the picture would still be the GPRS core network (now probably upgraded to EDGE or WCDMA), and the GGSN is the connection point to the service network. In this figure, the emphasis is on accessing the service network via a 2.5G/3G cellular system, but the connections to the IP backbone could also come from other access networks (WLANs, fixed LANs, circuit-switched 2G networks, and so on). Those networks are, however, out of the scope of this book—and we will not describe service network issues for those networks in more detail here.

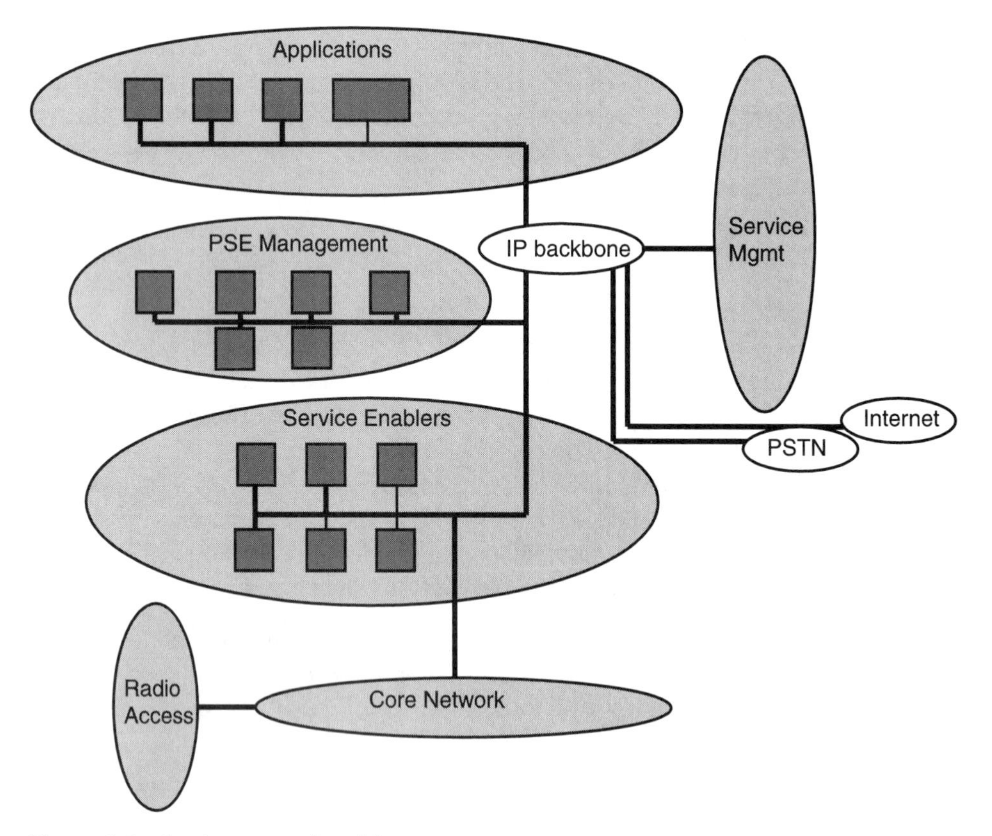

Figure 9.5 Service network architecture.

The architecture is component based, and the operator (here, the service network operator, which might or might not be the same as the mobile network operator) can choose which components to include and how. 3GPP standardizes the interfaces between components, and third-party developers can build components without having to sell the entire service network. For instance, a company might specialize in mobile positioning servers and sell that service separately to operators. More than likely, however, the company will sell the majority of service networks as packages that include installation and initial maintenance. Individual developers can still specialize in individual components and offer those together with the larger infrastructure players that sell complete solutions.

Another important aspect of the service network is the use of a three-tiered architecture for the service, as shown in Figure 9.6.

The functionality is divided between the different tiers in order make a more efficient use of resources. Most of the application functionality is located in the middle tier, where the application servers are designed to handle most of the calculations. The resource tier often consists of several components, including databases and different kinds of service enablers that provide access to the enhanced capabilities (as we describe next). You can then place each resource on one node or distribute them into several nodes across the network. Multiple applications can then access the same resource from the middle tier. The client tier includes the device that is being used and the presentation of the application. The presentation is, in this way, defined independently of both the application and the resources.

CORBA (Common Object Request Broker Architecture) is like a common thread throughout the service network and makes it easy for established developers to quickly gain an understanding of the inner workings of the components. CORBA has proven to be a flexible and high-performing standard, and it has the advantage of being commonly accepted already.

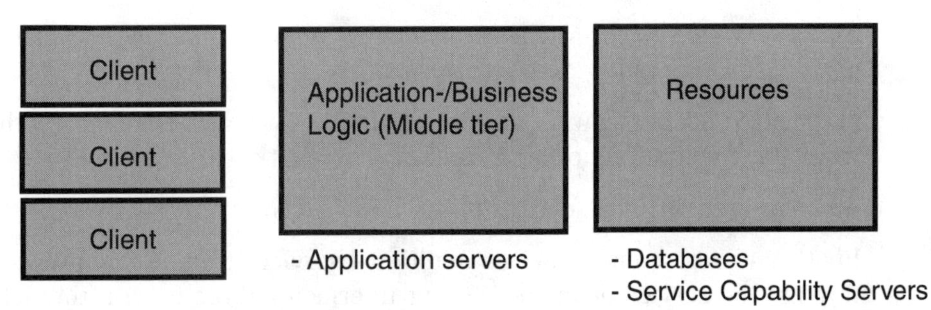

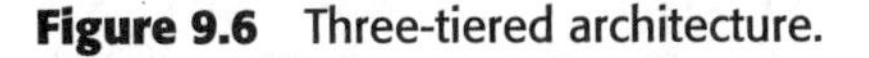
Figure 9.6 Three-tiered architecture.

We will now describe the four main areas in Figure 9.5 in more detail, examining the APIs and describing how applications developers can interact with the different components.

Service enablers. These are components that add new and (mostly) standardized services to the network. A mobile network, after all, only offers bearer services of different speeds and characteristics, and there is a need for standardized services on top of that. There are two kinds of service enablers: *Service Capability Servers* (SCS) and *Application Support Servers* (ASuS). The first kind, SCS, adds services that are built upon what the core network offers while the second adds services that are built upon other service network components. We will use some examples in the following section to highlight the differences.

Personal Service Environment (PSE). These components create a common set of personal preferences and profiles for users. Many applications can then reuse this information and give the user a more personalized service.

Applications. This area is where the applications servers reside, including application-specific databases and servlets.

Service management. As technology offers users more and more services, it is important to keep statistics for the users about usage and quality. In addition, it should be easy both for the operator and for the individual user to add or remove services dynamically.

Now, it should be obvious to you that a lot circulates around the service enablers.

Service Enablers

In Chapter 1, "Basic Concepts," we touched upon the definitions of applications and services and concluded the following:

End users consume services provided by applications, which operate on application servers and/or client devices.

In order to understand what service enablers do, we must complement the previous statement with another definition:

Applications consume services that service enablers provide.

In other words, applications get additional services from the mobile network through the service enablers without interfacing directly with the network. An example would be a positioning server that adds location-based services to applications. In the same way that programming libraries give convenient developer access to different features of a computer, the service enablers give access to features of mobile networks. The standardized APIs make it possible for an application to be compatible with a service enabler of a certain type, regardless of whether it is manufactured by Ericsson, Lucent, or somebody else. This functionality is one of the most widely anticipated features that software developers have pushed the telecommunications industry into adopting. Telecommunications networks have traditionally been very closed, and it has been difficult for third parties to get solutions into operation. The Internet model has so far not had this problem of inflexibility (but instead, a lack of structure and standardization of APIs). For example, it has been very difficult to get a micro payment standard to be accepted as a standard on the Internet.

Figure 9.7 is the same as Figure 9.5 but with the service enablers magnified.

The APIs not only make it possible to develop applications for all standardized service networks, but they also substantially raise the abstraction level. An applications developer will not need to know whether a user is equipped with a *Global Positioning System* (GPS) receiver or whether network-based positioning is available. A service enabler (here, a mobile positioning center or MPC) provides an API where the application can ask for the position of a user. The MPC then finds out the details by interfacing with the mobile network and with the device. If the application had interfaced directly with the network, you would also have needed to change it if you added more positioning options.

MPC is an example of a service capability server because it gives access to features that the core network provides, rather than to other service network components.

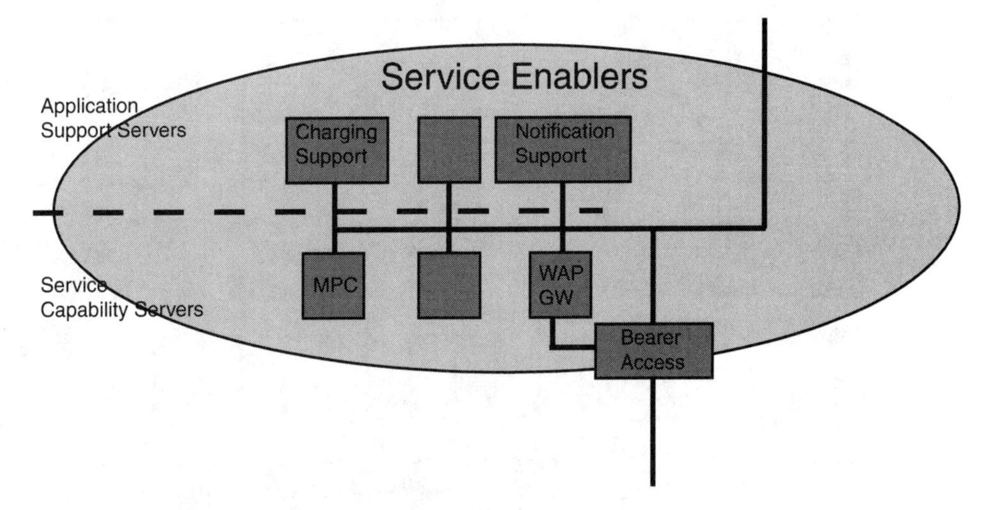

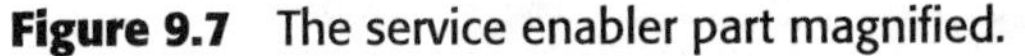
Figure 9.7 The service enabler part magnified.

Service Capability Servers

In order to make the VHE dream of personalized services for the masses come true; an open, standardized, and flexible architecture is necessary. As for most of the technologies that are primarily on 3G networks, 3GPP has performed this work. This time, however, another organization was formed in order to agree on APIs and protocols for service enablers. This group, called PARLAY, today consists of companies, and the group works closely with 3GPP and standardization ("Services to applications on application servers"). As we mentioned in the previous subchapter, a node (server) that offers such services that give access to core network features is called a service capability server. The services that such a server provides can be things such as user location, message transfer, and call control.

Figure 9.8 illustrates how the application server has access to a set of services that service capability servers offer, such as WAP gateway and gsmSCF (which

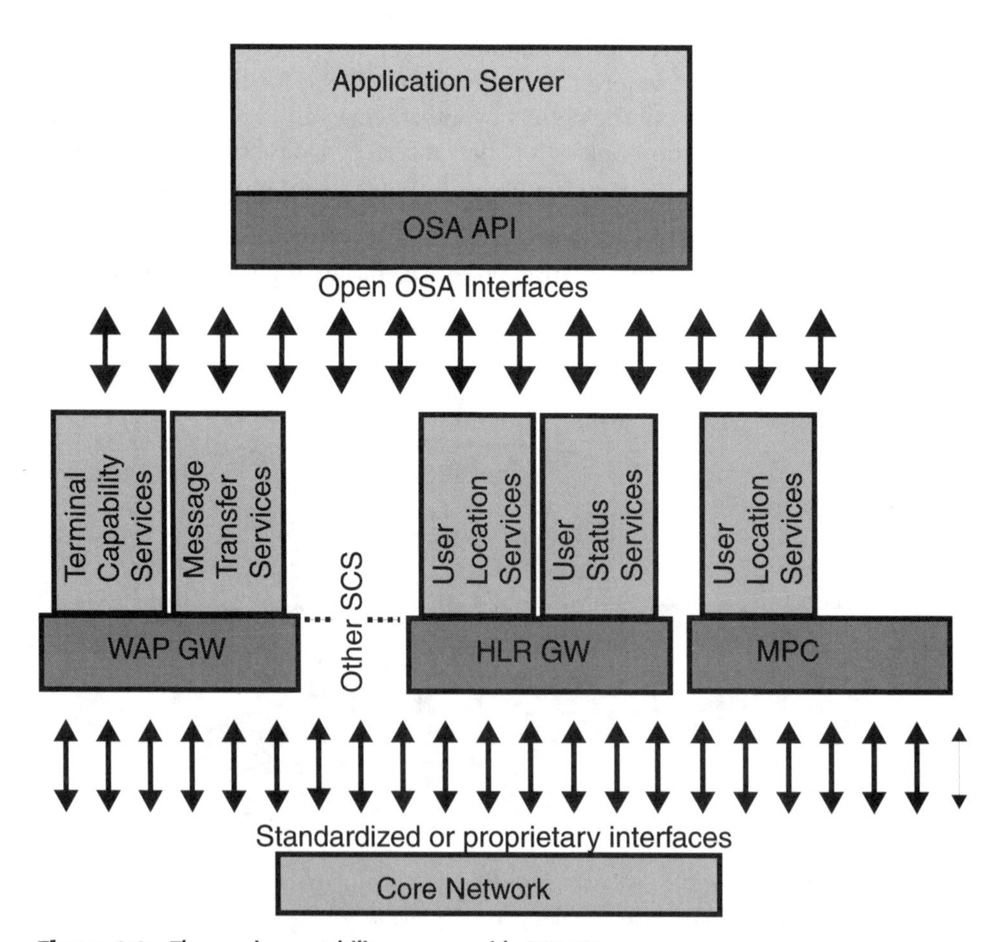

Figure 9.8 The service capability servers with OSAAPIs.

offers CAMEL Services (3GPP TS 22.078)). The architecture is sometimes called Open Services Architecture (OSA), and the APIs are all available in the 3GPP specifications (TS 22.078, TS 29.998).

As we see in figure 9.8, several SCSs can offer one service. The MPC as well as the *Home Location Register Gateway* (HLR-GW) offer user location services. Because the HLR is where the mobile system keeps subscriber information, it knows the cell in which the user currently exists. This information is then offered through the HLR-GW via the user location services. The MPC has this information as well, and it knows whether the user can be positioned more accurately with GPS or by using other technologies. In other words, the MPC (most of the time) can give a more exact position and thus a higher service to the application. Nodes such as the HLR-GW, which are not much more than an interface toward an existing node on the mobile network, are called OSA gateways.

Examples of SCS include:

gsmSCF. This node gives access to GSM-specific features.

WAP gateway. The WAP gateway, and WAP in whole, is actually a subpart of another service capability: MExE. MExE also includes Java, and it is described in more detail in Chapter 11, "Operating Systems and Application Environments."

HLR-GW. This gateway, which we mentioned previously, is an OSA gateway that provides access to HLR information, such as user location and status. The HLR is the node in the mobile network that keeps track of subscriber information.

Vendors of service networks are also likely to include some SCSs that have not yet been standardized and some that are standardized but are not included in OSA. One example is the *SIM Application Toolkit* (SAT) server.

Now we can see that all of these components offer services from the mobile network. All other service enablers are called application support servers.

Application Support Servers

The second half of the service enablers consists of applications support servers. The services that application support servers offer come from other service network components and external components (such as billing gateways). As with service capability servers, the application support servers can be implemented as part of a node that provides the service or as a gateway that interfaces one or several servers. We show the access of the services through APIs in Figure 9.9.

In the figure, there are two ASuS GWs that offer several services per gateway. The directory access support, on the other hand, is implemented as part of the common directory itself. The common directory holds subscriber information,

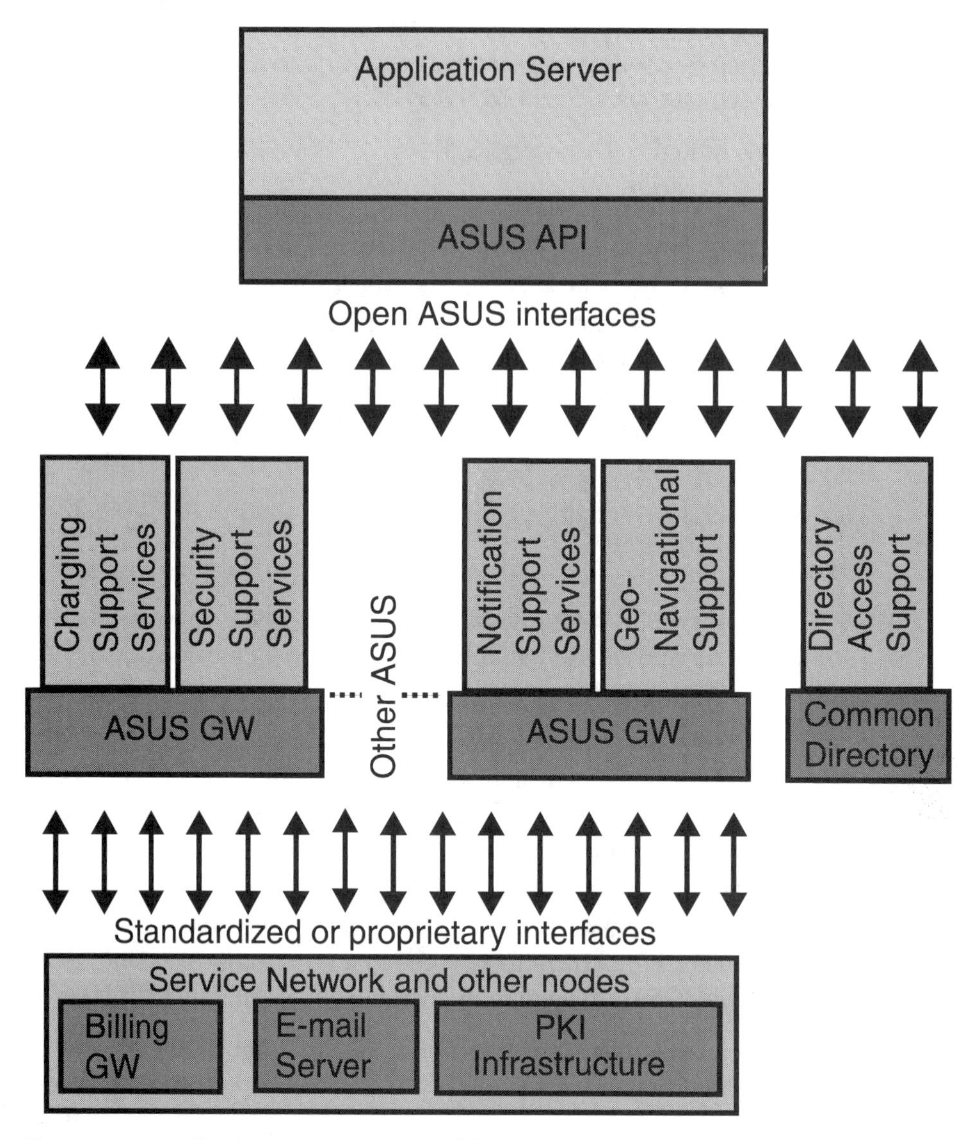

Figure 9.9 Application support servers with ASuS APIs.

and we describe it in more detail in the next section, "Personal Service Environment Management." Although these are two ways of implementing ASuS, the interfaces towards the applications remain the same. All of the ASuS support CORBA in order to facilitate platform and language independence.

The best way to understand ASUS is to work with it and to look at the different components and their interfaces. We give a few examples of ASuS here.

Charging support service. Mobile operators have great experience with charging customers for air time and services, but the advent of packet networks and IP applications creates some difficulties. Because the mobile system operates mostly on lower layers, it is hard to charge on a per-application basis with the built-in billing functionality. Some applications developers and operators are likely to use this feature in the future, however, because then they can charge users per game, per click, or per transaction. This functionality is now supported in the service network via the charging support service. This service provides a convenient API where an application can signal when and how the user should be charged. This information is then compiled into Charging Data Records CDRs (described in Chapter 3) that the billing system of the Mobile System understands. The operator can then chose whether this billing information will arrive to the subscriber on the same bill as the other 3G services or on a separate bill. He or she can also use the charging API to collect usage statistics and events.

Notification support service. Many applications are initiated by a notification from the server side of the application. This notification can be an e-mail alert that invites the user to start the e-mail client or an instant-messaging client indicating that a buddy has gone online. The actual method of the notification depends on either the user's preferences, as stored in the User Profile (part of the common directory) or explicitly stated by the application via the API. The choices of notification method range from SMS and WAP **PUSH** to network-initiated, circuit-switched calls.

Security support service. As more advanced applications emerge, it is crucial to have a powerful security toolkit at hand and APIs to support it. The security support service offers this feature and provides access to interfaces via the *Public-Key Infrastructure* (PKI) and other security enablers. Typical functionalities that are offered include certificate handling, authentication, and data encryption. We describe security issues and measures for applications in more detail in Chapter 12, "Security."

Geo-navigational support service. We already saw that positioning is a key component of future service networks in the form of service capability servers. Location-based services can involve much more than getting the position of a user, however (for example, finding the shortest route between two locations and finding things that are of interest in the proximity of a known location). The geo-navigational support service offers all of these services (which we predict will be some of the more popular ones). Location-based services is a large topic, and we will explore it further in Chapter 13, "Location-Based Services."

Directory access support service. As we will see in the next subchapter, there will be many more features available in order for the user to keep a profile and preferences that many applications can use. These services are part of the personal service environment (next subchapter) and thus are part of the service network. The directory access support service provides an API for the application developer to access this information. As for most ASuS, it provides an API for another part of the service network.

In order to use the individual APIs, please refer to standardization comments. You can find the detailed APIs in the 3GPP release '99 specifications (TS 29198-300 and TS 29998-300).

The open architecture opens new possibilities for third parties to develop their own SCS and ASuS nodes. Several opportunities are likely to emerge over time, and we will see products such as instant messaging servers, multiplayer gaming servers, and so on. The vendors of such servers will then release *software development kits* (SDKs) that developers can use to access the enhanced features. When examining these exciting new capabilities, we must obtain a feel for how widely supported they are. Which operators are using them, and how big is the market?

Personal Service Environment

Today, personalization on the fixed Internet takes place on a per-application basis with user registration and cookies. The first time you arrive at a Web site, you sign up to obtain a username and password. You sometimes also have to enter your personal information (in order to buy items, for instance) and preferences. The problem is that there is no connection between different Web sites, and you have to enter the same payment information for every *electronic commerce* (e-commerce) site that you use. Some sites use cookies to remember what you selected and what you are likely to want to see (in other words, your user preferences). Once again, the site that sets these cookies is the only Web site that uses them, and buying history books on Amazon.com will not help Barnes & Noble find out what kinds of books you like. We illustrate this issue in Figure 9.10.

In the figure, these three sites could potentially set a cookie that says that you like to read about ancient Greek history. You would probably find it nice if you could enter this information once and make it available to all applications (or at least to your three favorite applications). This concept is one of the thoughts behind adding the *Personal Service Environment* (PSE), where a common directory holds personalized information (see Figure 9.11).

This figure shows the thinking behind the PSE model, where information that is common to many applications can be stored in one place. This opens up new and useful features, like enabling users to log on once per session (rather than once per application) but still achieve personalized service.

While the service enablers do not hold much information themselves (instead, they facilitate access to data and services), the PSE does. The PSE is a set of

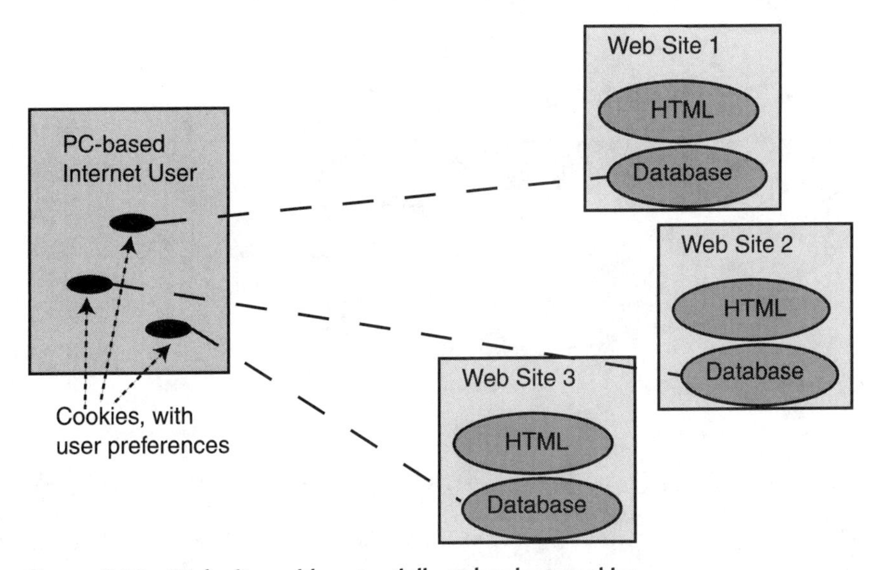

Figure 9.10 Web sites with potentially redundant cookies.

databases that hold subscriber information and preferences that you can access from a number of applications (see Figure 9.12). This feature does not prevent individual applications from maintaining application-specific subscriber data; rather, it is a complement. You should therefore use PSE for data that is generic and that is likely to be used by several applications (such as name, e-mail, favorite handset, and so on).

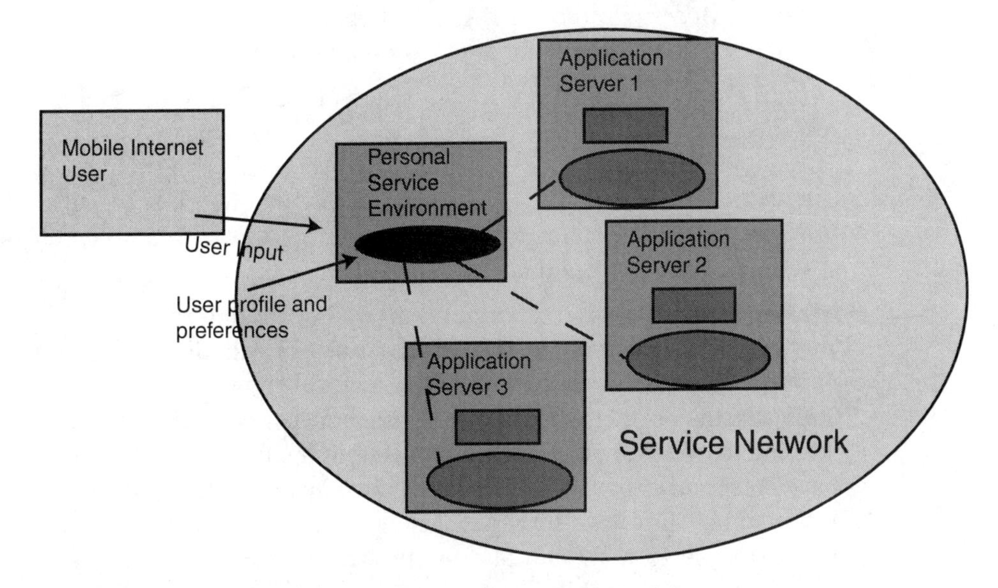

Figure 9.11 The *Personal Service Environment* (PSE) model.

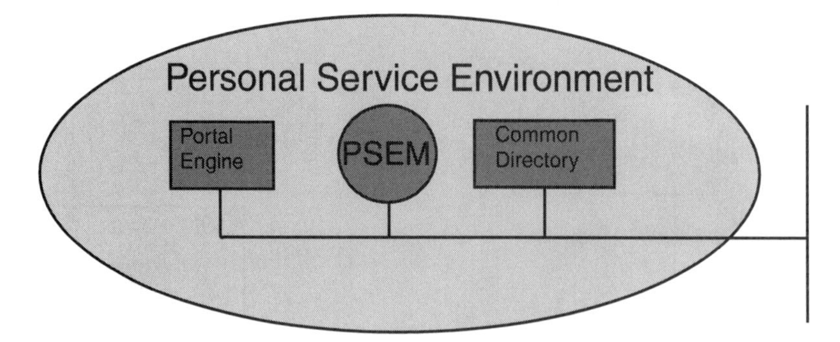

Figure 9.12 The PSE.

This figure is an extract of Figure 9.5 that shows the PSE in more detail.

The following paragraphs describe some components that can be part of the PSE.

Common directory system. This location is where we collect information from multiple applications and from the user and merge that data into a central directory. We divide the information into different groups, depending on who can access and change the data. The user can alter some parts of the data (personal address book, preferred services, and so on) while applications update other parts (maps, favorite actor, and so on). The User Profile is also part of the common directory and holds information such as name, e-mail address, services subscribed to, and so on. Information from the common directory is either accessed via an ASuS API (directory support service; see the previous subchapter) or via the *Lookup Directory Access Protocol* (LDAP). Either way, user preferences and profiles are now available to applications developers.

Personal Service Environment Manager (PSEM). PSEM's main responsibility is to manage the services to which the user subscribes. The PSEM enables users to see their subscribed services and their status. Users can access this feature through a WAP telephone or a PC that has an HTML browser. If the user wants to update any part of this common directory system, PSEM manages this procedure.

Portal engine. This general user interface manager determines how the services that are offered by the service network will look to the user. This portal engine also includes the service management that PSEM provides. Typically, the portal is the first thing that a mobile Internet user sees when starting the WAP browser. The services that appear on the portal can then be customized either by the WAP interface or on a PC. Typically, a user wants to use the larger screen and richer input mechanisms of a desktop PC in order to perform the initial customization and then to perform minor alterations on the mobile device (adding/removing bookmarks).

Application Servers

When you create an application, it is interesting to know the platform for which you should create the application. Generally, the application should be made as platform independent as possible. Discussing in concrete words how to deploy an application without focusing on a single platform would be impossible, and this book should be more generic than that. Every vendor of application servers provides (or should provide) an SDK that describes the exact procedure. While application servers theoretically could be made from any PC, there are some additional requirements that operators (and others who host applications) must consider.

The application server is the middle-tier part of the three-tier architecture in Figure 9.6 and can be defined as follows:

An application server is the place where the application/service logic for end-user applications resides and executes.

Consequently, the application server is the node where most of the actual computing burden is placed. Features usually include the following:

Linear scalability up to a certain load. In other words, you can buy the server with one processor board to serve 100,000 users and then add another board to double the capacity. At first glance, this situation might sound like a natural feature, but it is never achieved by a platform that is not optimized from the beginning. There have been examples of service providers that did not anticipate the rapid growth of mobile Internet services and failed to build a scalable platform for the application. In the end, you might add twice the equipment and only achieve a 20 percent increase in capacity. This scalability is further supported by the architecture of the service network, where load sharing between different servers makes it possible to scale the application server functionality even more.

Robustness and redundancy. One processing board can go down without affecting the performance of the other. The server can still be running while the faulty board is removed and replaced by a new one. This feature is obviously crucial when you consider that the service provider might now be a telecommunications operator who is used to having services and equipment that have a system down time of fewer than 10 minutes per year. The operator will not tolerate that the service that his or her customer experiences on the mobile is disrupted every once in a while because of server maintenance. The server must also be capable of handling a large number of hits without crashing, which is difficult to achieve but possible.

Zero down time software upgrades. If the server is serving a couple of millions of users, it cannot be rebooted if the applications on it are upgraded or installed. This situation requires a new way of thinking, because every second of service down time (when users cannot access the service) means a loss of revenue and customer dissatisfaction.

Multilanguage and platform support. While an operator or content provider obviously has to choose what application server to use, the application server must be capable of talking to service network components from other vendors as well. In other words, CORBA should be supported in order to enable the use of the standardized APIs (Parlay, 3GPP, and so on). For the convenience of the developer, the most common programming languages, such as C, C++, and Java, should be supported.

Manufacturers of application servers include IBM, Ericsson, and Sun Microsystems, to name a few. You can find more information about specific products on the developer's Web sites of those respective companies. Figure 9.13 shows an example of the exterior of an application server.

Application servers not only contain a run-time environment for applications, but also have certain middleware components that make development easier. These components can include object request brokers, transaction processing support, and message-oriented middleware.

Usually, the developer does not have to own an application server but can buy a hosting service from the service network provider (commonly the same as the network operator). We will see the advent of many new players in this field in upcoming years.

Service Management

As the complexity of the services adds up and more and more subscribers start using them, a good service management functionality is necessary. This functionality enables the owner of the service network to look at statistics and logs and to add and modify the user's portfolio. The more complex the solution becomes, the larger the need for an organized way to install new applications and update the service offering to individual users. In addition, it is valuable to be able to track the usage of individual applications and their geographical distribution. The live monitoring and changes to the configurations are usually done via Web interfaces in order to minimize the complexity for maintenance personnel.

The service network management functionality is out of the scope of most applications developers. The owner of the service network is the one who maintains and handles this feature.

Figure 9.13 An application server's main cabinet.

Summary

Traditionally, the mobile networks have been closed, and it would have been hard for third-party application developers to create products for them. Applications were designed with one specific network in mind, and it was difficult to create an application that could run over several bearers. One of the most important aspects of the 3G networks is that the architecture now is layered horizontally. In other words, applications can be designed to run independently of the underlying networks and still interface with the control and transport functions. The open APIs that facilitate this process are specified by Parlay and 3GPP and make it possible to add features to the applications (such as positioning, call control, and personalization).

CHAPTER 10

Mobile Internet Devices

One of the biggest open issues for developers of wireless applications is that no one wants to tell them what future devices will look like and what accessories and properties they will have. Because many software developers are used to working with hardware developers who show hardware years in advance, this transition is difficult. One example is the people from the interactive entertainment and gaming industry. Sony, Nintendo, and Sega have been very generous to their closest developer affiliates for many years and have sent detailed descriptions and *software development kits* (SDKs) long ahead of their commercial launch. This relationship is a must, because it takes quite some time to get used to a new platform and to learn to develop efficiently for it. Consequently, there has been a very high correlation between the success of a hardware/software platform and the amount of available software. Mobile telephones have traditionally been closed platforms with just a few applications per device. Manufacturers of mobile phones and PDAs are not accustomed to this kind of openness that the software industry is used to, and both sides probably have to adjust.

What then is the information that developers need in order to be able to get great applications out on the market quickly? The tricky part is that the fresh information is of the highest importance to developers (for example, what are the properties of devices that will come out one year from now?), but this is also the most confidential information. When we describe the properties of the devices of today and tomorrow, we will use a generic approach that will apply even years after this book's publication. As a result, the developer will be better positioned toward making decisions about future products, and it will be easier to use the up-to-date information that manufacturers provide through SDKs.

The question is, "How much does a wireless application developer need to know about upcoming devices?"

Devices Now and in the Future

In fall 2000, I attended a panel discussion in San Francisco about the future of mobile devices. The audience consisted of designers who, at that time, mostly worked with Web design and publishing on desktop PCs and who were interested in designing for the mobile Internet. I could see the crowd becoming increasingly weary as we talked about the devices of the future and how the form factor would evolve but would still be limited. Almost everyone could agree that the mobility would always limit the size of the device (until we invent screens that you can fold like maps), so that point was acceptable to the crowd. Then, someone wanted a bit of comforting from me and asked whether we are working on a way to standardize the screen size and other output properties in order to make life easier for designers. I felt like I kicked someone who was already lying on the ground when I said that things will only become worse for designers in the future. The more features that become available, such as high-speed access, more powerful hardware, and better screens, the more diverse the mobile devices will be. There will be no "one-size-fits-all" device that has all the features available and still is the smallest and most power efficient.

The industry always tries to standardize the things that it can standardize, but there are many components of these devices that it cannot standardize fully. The closest that we come to standardization is the MExE framework (described later in this chapter), which enables devices to divide into different classes depending on their capabilities. In addition, platform developers like Symbian have developed frameworks (Quartz, Pearl, and so on) for device families. Using those frameworks, you can develop for several devices that use the same operating system and device family. The application should not despair, because there are ways to design applications that will work on multiple devices (as we will discuss later in this chapter).

In the evolution of devices, we can see two tracks that evolve in parallel: horizontal devices and vertical devices.

Horizontal devices often focus on the mass market; therefore, making them cost efficient is of high importance. You must also conform to all possible standards in order to make application development easy. SDKs for horizontal devices are often distributed via the manufacturer's Web site and are often available to anyone who is interested in supporting them. Horizontal devices are not primarily designed for a specific application or

Figure 10.1 Horizontal devices.

usage model. Examples of mass-market devices include the Palm V, Nokia 3310, and Ericsson R380 (see Figure 10.1).

Vertical devices fulfill a special need for a vertical segment. In other words, devices are generally more expensive and manufacturers do not have to have manufacturing costs as their top priority (which does not mean that cost is not an issue). SDKs are sometimes publicly available, but developers often have to be in a closer relationship with the device manufacturer or with the one who makes the operating system/middleware on which applications are developed. The big advantage of developing for vertical devices is that hardware and software are made specifically with certain uses in mind. A mobile game console, for instance, can have a small joystick or a four-directional keypad for maneuvering. Examples of vertical devices include a Symbol device with a bar-code reader (see Figure 10.2), mobile game consoles, and in-car mobile clients.

The variation between different devices is generally more noticeable within vertical devices, and the provider of the software platform often provides

Figure 10.2 A bar-code scanner from Symbol.

enough tools to get into development easily. For this reason, most of the focus of this book is on horizontal devices. First, we will look at some of the factors that affect future devices and the challenges that the manufacturers face.

Building the Ultimate Device

Many of us have probably thought that it would be great to create the ultimate device with all of those features and a nice design (okay, I guess that puts us in the "geek" category). Not many of us realize what it takes to build a device (and more importantly, make the device a profitable success). The aim here is to narrow the gap between device manufacturers and software developers by making each side understand the needs and challenges of the other party. Note that we use the word *device* throughout this chapter, because it includes everything from WAP phones and *Personal Digital Assistants* (PDAs) to in-car mobile terminals and wirelessly enabled soda machines.

Business Aspects

The traditional way of selling mobile phones is together with a mobile subscription. In other words, manufacturers such as Ericsson, Motorola, and Nokia sell the phones to the mobile operator that owns the network. The operator then sells a network access subscription along with the phone. One part of the phone sales is together with network infrastructure sales, which often gives the operator a chance to put requirements on the phones that are to be offered. Therefore, many mobile phone manufactures cannot make some decisions about upcoming products without taking the operators' needs into concern. If the operator is happy with the phones that the manufacturer offers, the manufacturer will market these phones in media campaigns and in retail stores. Some operators also go one step further and brand the phones. Sprint PCS in the United States, as well as NTT DoCoMo in Japan, are examples of the operator reselling phones with the operator's name on it, although someone else manufactured the main part of the phone. If the subscriber signs up for a subscription that spans a year or two, the operator usually subsidizes the phone and a brand new phone can then cost as little as $1 (Euro) or less. The operator then counts on getting the money back from the call charges that the user acquires during the subscription period. This is of course a model that operators want to move away from, and in some markets we are already seeing fewer phones being subsidized.

Now that PDAs and hand-held computers are becoming increasingly wireless, some of those devices will join this model as well. This group has traditionally been resold as any electronics equipment through departments stores and electronics warehouses. Those devices that become wirelessly enabled via Bluetooth can easily keep this distribution method, however, because no subscription is needed for usage. Those who have built-in GPRS and 3G wireless functionalities are a bit more complicated, however, because a subscription from a mobile operator is necessary in order to use wireless properties. Possibly, these will be sold just as mobile telephones always have been—via the operator. Of course, there is also nothing stopping an electronics warehouse from reselling mobile operator-subsidized PDAs with 3G subscriptions.

One important aspect of the operator's involvement is that the device must not limit the choice of business model. The operator's portal and service network components must be configurable and given a prominent placing. This process requires many options to be configurable, which in turn can potentially make the device less user friendly. Many users will get their first mobile Internet device after only using a cellular telephone, where they did not have to do anything more than dial the number and press the dial button. You cannot expect those users to pull up the configuration menu and choose what WAP gateway to use. One solution is *Over-the-Air* (OTA) configuration, where a message (usually an SMS) that contains the settings is sent to the phone from the operator. The phone then interprets this message and automatically configures the device accordingly. Nokia and Ericsson phones support this feature, but there is no standard available.

The Man-Machine Interface

Even if advances in battery and display technology result in larger displays with color support and a higher refresh rate, the size of those displays will still be limited for the majority of devices. For hand-held devices, the name indicates that users are not likely to want devices that are significantly larger than today's PDAs. The task for device manufacturers is instead to get as much display from the limited form factor without sacrificing battery life. The display of a color PDA stands for a large part of the total battery consumption, and a larger display means a shorter battery life. Another factor that limits the display size is also the robustness of the device. A large screen is more sensitive than a small one. I have dropped my phone on the floor many times without any problems, while my PDA broke the first time that I dropped it on the floor from a distance of one meter.

By saying *Man-Machine Interface* (MMI) we mean the way in which the device enables the user to interact with it. The MMI is constructed so that it enables the user to control the device in sufficient detail without making things too complicated. Typically, the MMI makes choices by asking the user questions, which removes some of the details that are likely to cause confusion. For example, a user might want to connect his or her laptop to a *General Packet Radio Services* (GPRS) network via a GPRS-enabled PC card. The modem software on the laptop then gives the user easy access to the network card, and one click on an icon initiates the connection. In the background, this action has Attached the GPRS user to the network, initiated a *Packet Data Protocol* (PDP) context activation, and obtained an IP address. A hand-held device also needs to provide this kind of abstraction.

The main challenge of designing the MMI is the large number of features that many devices now have. The R520, for example, has five bearers over which a connection can be made: GSM circuit-switched (HSCSD), SMS, infrared, GPRS, and Bluetooth. For future devices, you can add EDGE and WCDMA to that list. If the user then wants to use the WAP browser, how can the device know whether the user wants to use SMS, GPRS, WCDMA, or HSCSD in order to attach to a network server (or perhaps use WAP over Bluetooth to access an information kiosk that is nearby)? More than likely, the majority of users will not be able to (or want to) make these kinds of decisions.

One solution is to preconfigure a preferred bearer for each kind of service and then let this bearer be configurable through settings on the device. For sending business cards, it is likely that the preferred bearer is Bluetooth (or possibly infrared), while GPRS conveniently handles WAP. In this way, the MMI makes it easy for the user to get started with the device, but advanced users can still affect the configurations.

One of the great challenges of future devices is to handle the different *Quality of Service* (QoS) classes of (primarily) the 3G systems. With the different parameters of each QoS class, there will be tens of thousands of QoS combinations. If the operator then wants the users to pay different prices for different levels of QoS, things become tough. Some of the QoS settings will be available to the application through APIs (not yet specified). This point is where application developers will be even more closely involved with the mobile Internet value chain. Suddenly, the way in which the application selects these QoS parameters affects how much money the subscriber is charged. This situation cries out for a closer relationship between the involved parties.

Communication

In addition to the dilemma of knowing which bearer the user prefers, numerous other issues are connected to the design of wireless devices. Wireless standards are generally constructed so that the network can choose whether to support some features that can then be added later as a network is upgraded every once in a while. The terminals and the devices, on the other hand, must usually support a majority of the functionalities from the beginning. One example is the coding schemes for GPRS. As we described in Chapter 3, "GPRS-Wireless Packet Data," there are four coding schemes (CS1-4) while some infrastructure vendors will only support CS-1 and CS-2 initially. Handset manufacturers, on the other hand, have to support all four coding schemes.

In general, the devices are also harder to upgrade, and you cannot take it for granted that users are willing to upgrade. Some devices are starting to support upgrades that are sent over the air, but this situation is a new and strange concept to most users. This constraint implies that device manufacturers need to get everything right before releasing a product.

Interoperability is another issue, and devices need to be tested based on infrastructures from different vendors in order to ensure operation on all networks that conform to the standard. With the advent of Bluetooth, the communication with other Bluetooth phones, Bluetooth headsets, and Bluetooth-enabled laptops also needs to be verified. The devices that have an open operating system should, of course, also be tested against the most popular software applications that run on the platform. Add to this situation that the WAP Forum now certifies devices, and you have lots of interoperability testing that is needed for the average multifeature device. We combine these factors with living in a world that gets more and more influences from the fast-paced software industry.

Hardware Advances

Moore's Law, which says that computing power doubles every 18 months, has been a fundamental driving force in the PC industry. The hardware advances within that segment have also opened new possibilities for multimedia applications such as games, streaming music, and video over the Internet. Now that the mobile Internet has started to deliver higher bit rates and more functionalities, people are starting to speculate when the richer user experiences will reach the mobile devices. One limitation is, of course, the applications environments and operating systems (which we will describe in the next chapter). These environments have to become more open in order to enable the richer content. More importantly, the devices must have hardware that supports the more advanced applications.

Measuring the power of the CPU in a mobile device is a bit different from desktop PCs because it is more interesting to look at the power consumption together with performance than to examine the computing power alone. This value is commonly measured in *Million Instructions per Second per Watt* (MIPS/W). This unit describes how much energy each instruction in the CPU consumes. In the year 2000, a common figure is 2MIPS/W for a 200MHz processor and 1.25MIPS/W at 400MHz. The increase over the past few years indicates a similar development as for the Intel-based PC processors, where speeds increase exponentially and power consumption still goes down.

In a similar development, the power consumption of the displays is reduced drastically. In 1998, 120mW was a common power consumption level for a 1/8-inch VGA screen (Palm V size). This value was reduced to 8mW in 2000, which significantly increased the battery life. Battery-life advances still occur, but not in the same pace as the rest of the electronics industry (such as CPUs).

As the need for multimedia now emerges, more and more dedicated chip solutions are popping up. These solutions include graphics (for example, .GIFs and .JPEGs), sound (such as MP3s), and video (including MPEG-4 and MPEG-7) compression/decompression, GPS receivers for higher positioning accuracy, and also some more specialized chips (such as 3D graphics). For vertical devices such as gaming consoles, it is easier to foresee upcoming needs. Because many of the chips that are now appearing are already significantly developed for PCs, manufacturers might have the advantage of learning from those experiences. On the other hand, building a chip for a mobile device is significantly more difficult because of issues such as miniaturization and low power consumption requirements. This situation might be a good opportunity for new players to enter this space. Even more difficult than those issues is likely the judgment of experiences from the PC world that are applicable to the mobile world.

Input Mechanisms

With the advent of the computer mouse together with the *Graphical User Interface* (GUI), the user friendliness of PCs took a quantum leap. The interesting thing was that these technologies did not make the old input mechanism—the keyboard—superficial; rather, they complemented it. Still today, there is no faster way to produce a digital document than to write it on a PC by using a keyboard. Both the keyboard and the mouse fit the mobile world poorly because they both require lots of space and a supporting desk, table, or lap. Mobile phones come from the other end of the spectrum and provide minimal input mechanisms (a keypad plus directional buttons/wheels). Most 2G phones were designed to be operated with one hand and were sometimes complemented with voice control.

Voice Control

We can divide voice control into two types:

- Handset based, where the processing of the command takes place in the device (for instance, "Call Susan."). This feature is ideal for short commands that are easy to interpret, and the commands are often tailored to the user of the device. The user has to program these commands by recording them, and then the device compares the recording to the command when it is used. The main advantage of this approach is that the response time is very quick, and users feel like the device instantly acts on command. The drawback is that this method cannot easily be used for more complex commands and commands other than those that were previously recorded. For commands that are not previously recorded, the device needs to recognize what words the user is using, and this process requires substantial computing power (and therefore consumes lots of battery power). For those applications, we prefer a network-based solution.

- Network-based voice recognition is usually implemented on a separate server node. In other words, there is lots of computing power (and maybe even dedicated hardware) that can interpret the user commands and output the meaning to an application. This translation from a spoken word into a digital command that applications can understand and act on is already used in some travel booking services. The user can call a number on his or her telephone (regular or mobile) and become connected to an application server that performs voice recognition. Even advanced commands such as, "I want to travel to Boston next Friday." can be interpreted by some systems. The main drawback of this system is that there is a delay between the user's command and the response from the system, because they are often located far apart. Even if the server were located in the operator's service network (adding a service capability), the user would perceive a delay that could be annoying for highly interactive applications. Figure 10.3 shows two examples of where you can place voice-recognition servers.

While voice call access will still be useful, voice-recognition servers open new and exciting applications with the advent of packet data. A device can record a voice recording, package the data into IP packets, and send it to an application server. This server can then choose to interpret this information by using voice-recognition technology, or the actual voice recording can be stored (team database of voice memos) or sent to other users. Applications include workforce coordination and instant messaging.

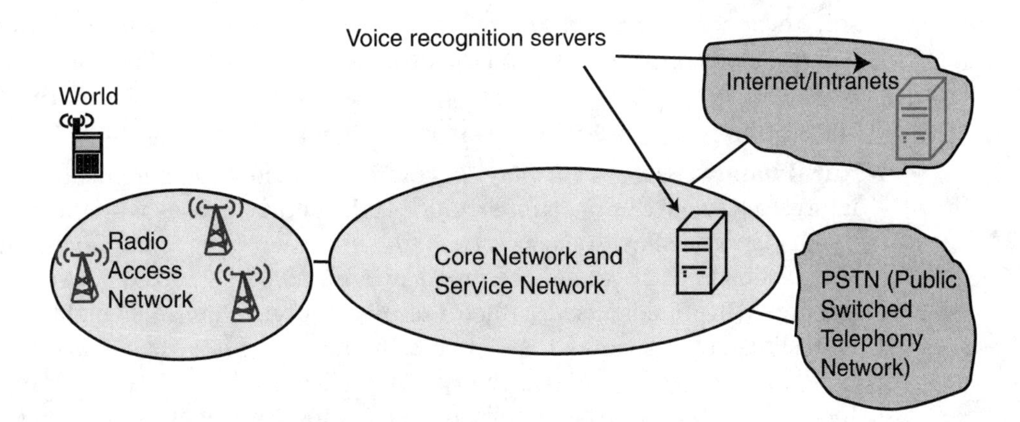

Figure 10.3 Example voice-recognition server locations.

In order to standardize the use of voice commands for browsing, a standard has been developed called VoiceXML, a standardized successor of VoxML (www.voicexml.org). This technology enables the construction of sites through which users can navigate by using voice commands. The site construction markup language is similar to HTML/WML and is fairly easy to manipulate.

Pen-Based Input

The breakthrough of the PDAs in the mid-'90s, along with the Palm OS-based hand-helds that were on the cutting edge, made users accustomed to using pen-based user interfaces. Different devices have different approaches to this feature, however, as we will describe next.

Character recognition. The more intuitive way of constructing a pen-based device is perhaps to try to identify what letters the user writes while enabling the user to write as if on paper. Microsoft Pocket PC (Windows CE)-based devices use this method, which usually does not require a long learning period. The disadvantage is that some advanced users find the characters complicated to write, and it is easy to make mistakes. This application also requires quite a lot of CPU power on behalf of the device.

Simplified character set (Graffiti). The Palm platform became popular in the late '90s, and one of the key features of that platform has been the Graffiti technology. With Graffiti, the character-recognition process is simplified by a special character set that makes characters easy to recognize and separate. This process makes writing quicker (after an initial learning period), and the clearer distinction between characters generates fewer mistakes.

Virtual keyboard. On most devices that have some kind of character recognition, there is also a virtual keyboard that you can pull up on the screen. Others, such as the R380, bases most of the character input on a soft keyboard on the bottom part of the screen. The learning phase for this technology is very short, but on the down side, it is hard to become substantially quicker with it. Typing large amounts of text is therefore not appropriate, but that probably applies to most pen-based technologies.

Natural handwriting. As displays become bigger and better, it becomes interesting to have applications that read the actual strokes with the pen on the display and show them as they were performed. This feature opens up the possibility of free-hand chat applications and personalized notes and pictures. The personalization factor of this input mechanism is highly appealing, but it also has some obvious limitations. First of all, it is difficult for computers and devices to interpret information that exists in a drawing format or as text with personalized handwriting. Second, the message that results from a scribble on a screen is really a screen capture, which is a pretty large file even on a small display. This functionality requires an advanced compression technology. Nevertheless, there will be some very appealing applications emerging from natural handwriting techniques.

Pen on paper. Finally, there is the possibility of using a Bluetooth-enabled pen that reads the pattern of a piece of paper. The paper has a pattern that makes it possible for the pen to know its location on the paper via a small camera, which is located at the top of the pen. The pen can then send this information back to the phone/PDA/laptop and reproduce what the user wrote or drew. Anoto (www.anoto.com) is an example of a company that has such a product, and it will be interesting to see how applications developers will develop complementing software for this technology (see Figure 10.4).

The Keyboard

For laptops and some hand-held computers, it is likely that the keyboard will remain the main input mechanism for text. When it comes to typing large text segments, the keyboard is hard to beat, at least when using the western alphabet. With the advent of Bluetooth, some of these keyboard-based devices (including desktop PCs) will serve as input devices for applications that run on smaller devices with more restricted input mechanisms. For example, you can more easily update and organize a large address book with a laptop and a keyboard than you can with a phone that has a keypad. Most PDAs have a PC-based user interface as well and leverage the input friendliness of the PC. Because Bluetooth enables communication with virtually any device, this functionality will extend into more areas. Perhaps a Bluetooth-enabled laptop or hand-held computer will set the labels on individual tracks of a Mini-disc player?

Keypad and Other Input Mechanisms

On small, phone-centric devices, it is likely that the numeric keypad will remain—either as the real buttons that 2G phones have, or as a soft keypad on a touch screen. The keypad is very intuitive for anyone who has used a regular phone, which makes it the choice for first-time buyers. Whenever you use the keypad, you should complement it with something that enables easy navigation. The biggest problem with the keypad is that it is slow for entering large chunks of text. The use of predictive text input, such as T9 from Tegic (now a part of America Online, or AOL), helps a lot, but still other mechanisms are preferred for writing e-mail and for using other applications that require massive input.

Figure 10.4 Anoto technology with a Bluetooth-enabled pen.

Device manufacturers have already experimented with wheels, rollers, and multidirectional pads in order to enable easier navigation. There will be more innovation in this area, with everything from joysticks to motion detection, and the developers can only wait and see what the innovators will develop.

Handling Multiple Input Mechanisms

Even with 2G phones, there is commonly more than one way of inputting information into a device. We will use this functionality even more in 3G devices, where you can combine voice recognition with pen-based input for some devices (and use other combinations for other devices). The starting point for applications developers should always be to take nothing for granted unless absolutely necessary. Then, when you cannot add any more functionality without assuming a certain input mechanism, you should isolate this functionality as much as possible.

Integrated versus Divided Concepts

When more and more technologies appear in the mobile Internet, it becomes harder and harder to construct a one-size-fits-all device. Some people do not want to add a single gram of weight to their phone just to get some fancy feature. Others, however, dream of the device that has it all—all of the communications technologies, the fastest processor, and the biggest screen. Even for individual users,there is a big difference between preferences during a normal week. Perhaps the super communicator that weighs hundreds of grams with everything in it is perfect as long as you have your briefcase with you. The next day, you might be walking to a restaurant at night, and the tiny phone that fits into your pocket is ideal in that situation. There will likely be a need for all sorts of form factors, and you can divide the usage models into two categories: the *integrated* and the *divided concept*.

The integrated concept is the device that combines the modem with the application environment. The thought is that the user will need only this device and nothing else (that it should be both a phone and whatever other functionality the user needs). Figure 10.5 shows some examples and illustrates two smartphones. Communicators such as the Nokia Communicator and PDAs with built-in voice functionality also belong to this category. With all of these devices, the modem communication is on the same physical device as the application executes. Advantages of the integrated concept include the following:

Setting up and preconfiguring the system is easier for the operator. With one device, it is easy to send a message (for example, SMS) to the device with the settings, such as WAP gateway address and DNS and proxy details.

If something is not working, there is only one device that you have to examine in order to find the problem. A common problem for those who buy components of a desktop PC and build it themselves appears when something goes wrong. The guy who sold the motherboard blames the guy who sold the memory chip. The same issue might appear for those who buy a phone from an operator and a PDA from a nearby electronics store but avoid the integrated concept. The device is bought from the operator or the store, and the one who sells it feels responsible for helping make it work.

The application developer will know what access method you can use and can design applications accordingly. There is a significant

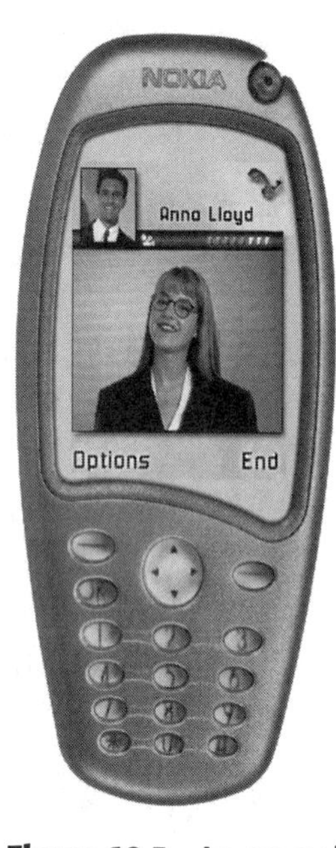

Figure 10.5 An example of integrated concept devices.

advantage if the developer knows that the device will have packet data access (for example, GPRS) or if he or she knows that a certain degree of QoS will be available (such as UMTS).

The application developer can see the entire picture and will know more what the user will experience. If someone designs for a high-end communicator, it is easier to foresee the potential market for the application and also the user's behavior.

The operator can anticipate usage and better plan for which applications to offer and how to dimension the network. With the entire functionality in one package, the operator can better know the requirements for bandwidth and the service network needs (positioning, multimedia needs, and so on). This knowledge helps the operator not only dimension the service network properly but also do a better job with radio planning (determining how many base stations and where to place them).

There is a good opportunity for operator branding. The application runs on the same device as the modem, which means that it is a communications device with rich applications functionality. Operators who want to position themselves as leaders in applications and not just in transporting bits and bytes are likely to find a strong appeal with this opportunity.

Disadvantages of the integrated concept include the following:

Low flexibility. If the modem part becomes outdated, the entire device needs to be replaced. In a world where more and more functionality is quickly added to the mobile networks, it is hard for device manufacturers to anticipate what the customer demand will be when a device is launched. This situation is especially true now because it takes two to three years to develop a mobile phone (and even more time to develop complicated communicators).

More cumbersome. Due to the large amount of features, it may be larger and heavier.

The device is a package of features, and it is unlikely that those specific features are the best ones for all situations. The device manufacturer has to make choices of what to include and what to not include in order to market the product in time. The resulting features of a device become the average of what the target customer group is interested in, which might feel like a set of compromises for some. With the advent of more devices that have open platforms, this issue will no longer be as prominent. At least, the software applications can be upgraded.

For the application developer, it is harder to calculate the CPU power that is available and the battery drain level. This is because the communication part heavily affects these factors, as well. In a typical communicator, a lot of system resources are spent on the communication part, and for packet data devices, the usage varies greatly. This situation gives the application developer more things to consider.

We show the divided concept in Figure 10.6, where the *Mobile Terminal* (MT) and *Terminal Equipment* (TE) are physically separated. The R520 is a Bluetooth-enabled GPRS handset that here only acts as a GPRS modem for the Bluetooth-enabled PDA (note that the PDA in the figure is not Bluetooth enabled but merely illustrates the concept). When the phone and the PDA are paired (see Chapter 5 for Bluetooth details), the PDA will have a constant connection to the GPRS network (and probably to the Internet as well). An application on the PDA can now act as if the PDA itself had a network connection and an IP address. You can also use the divided concept with 2G phones, where a cable or infrared connection primarily connects laptops to networks.

Advantages of the divided concept include the following:

The TE can be a device that is optimized for applications, and you can optimize the MT for modem usage. In other words, the functionality for each device is tailor-made for the task, and you have to make fewer compromises.

High flexibility. If one of the components becomes outdated, you can easily replace it. While people tend to use PDAs and laptop computers for several

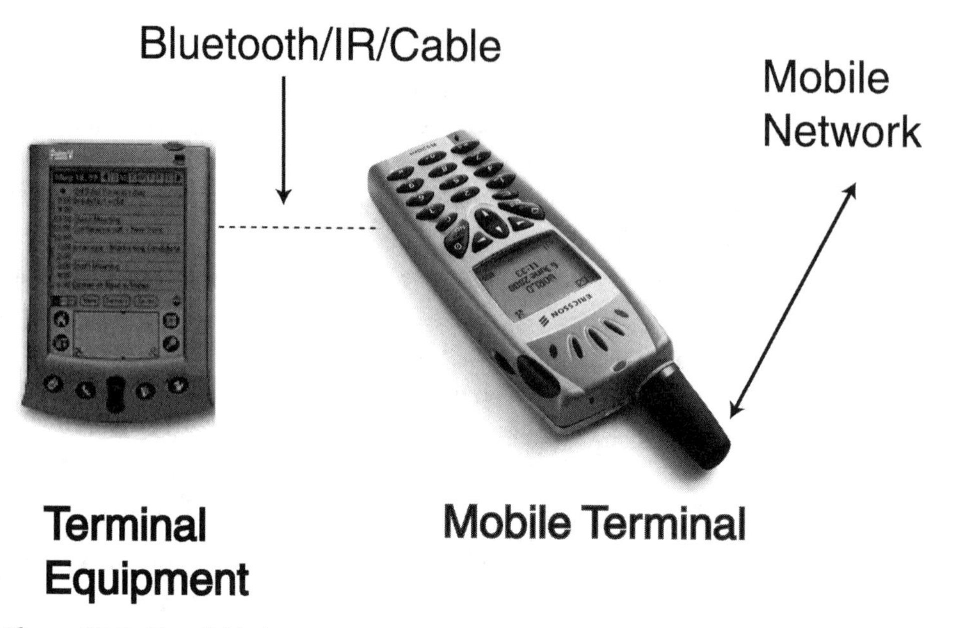

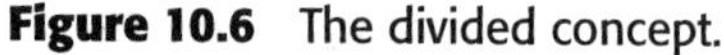

Figure 10.6 The divided concept.

years, they replace phones more quickly. Keeping most personalized details in a device that has a longer life span is an advantage, because it keeps the phone simple and easy to replace.

You can use the same MT (modem) with several devices that all are specialized in different applications. In other words, you can buy one phone that you can use both with your Bluetooth-enabled laptop and your PDA. Maybe there is even a need for two PDAs, where one is the ultra-slim organizer and one is the feature-rich multimedia device.

Disadvantages include the following:

If you do not use Bluetooth, the physical usage can be very cumbersome. Infrared connections require line of sight, and cables tend to always find innovative ways to create knots and produce a general mess. Even with Bluetooth, you have to make sure that the devices are in range in order to work together.

The configuration is more complicated. The phone (MT) needs to be configured in order to talk to the mobile network (for example, GPRS) *and* to the TE (the PDA or laptop). In addition, the TE needs to be configured accordingly. Enabling OTA configurations is also difficult for the operator because the devices are physically separated.

As we can see, there is no straight answer to what the optimal solution is, and we are likely to see many combinations in upcoming years. The application developer is put in a situation where not only the variety of devices increases rapidly, but also the number of combinations that can be created with the divided concept. In an attempt to classify different devices into groups that developers can design for, MExE has been developed. We describe MExE in the next chapter about application environments.

Designing for Generic Devices

We have shown that there will be an even bigger span in functionality, display size, and wireless access possibilities. Now, what will we take for granted, and how can developers accommodate these possibilities? Although there is no straight answer to these questions, there are some generic guidelines that we can use in order to make the support of multiple devices less painful. As we described previously, one of the keys is to isolate all device-dependent design issues into one part of the application that we can update independently of the rest. If the application is designed as a browser-based, client-server site, the key is *Extensible Markup Language* (XML). As Chapter 7, "The Wireless Appolication Protocol (WAP)," describes, XML enables you to clearly separate the content from the templates that guide the presentation. For those applications that are written in a programming language and run on the device, a good way is to make the application highly modular and object oriented, but there is no generic miracle medicine (although I would hesitate to call XML that, as well).

One of the key concerns (regardless of platform) is to save battery life. No one wants to be the designer of an application that drains the battery of the users' devices. Most of the remedies come from common sense. The display, the CPU, and the communications parts are likely to be the main power consumers in the device, and you should always give thought to how you could optimize the usage of those resources. Event-based programming is advantageous here, because we can use it to help avoid the infamous idle wait loops (repeating the same code thousands of times until something happens). Usually, the event-based approach enables the CPU to rest when nothing is happening. With the communications part, the key is planning. Being always online does not mean always talking to the server side. Not only does this situation leave the application exposed to delays, as we described in Chapter 8, "Adapting for Wireless Challenges," but you also consume power when you turn the transmitter on and off all of the time.

Although the form factor of devices will be uncertain one year ahead of whatever date you pick, there are some things that you can count on. One thing is that the majority of users will want to have a device that is small enough to hold in one hand.

Some of the aspects that affect the design of applications depend more on the operating system and applications environment on the device, and those things tie tightly into what we have described in this chapter.

Summary

The applications highly depend on the devices that are available, so we expect to see many new and exciting models for the 3G systems. The only thing that we can take for granted is that the diversity among devices will only become bigger. They will vary in size, shape, and functionality. Although we are constantly making progress with creating more user-friendly devices, the input mechanisms and battery power are likely to be limiting factors that we should consider.

CHAPTER 11

Operating Systems and Application Environments

In the first part of this book, we looked at the new networks and radio technologies that bring a multitude of new services and enablers. Together with the rich spectrum of new devices that we described in the previous chapter, there is now one major piece of the puzzle that we need to explore: the environment and platform on which the applications should execute. We touched upon this topic earlier when we looked at the *Wireless Application Protocol* (WAP) and how it copes with the difficulties of the wireless networks, but WAP is only one way of creating the applications of the future. As the devices become more powerful, there will be a whole range of them that offers some kind of applications environment with varying degrees of freedom. Some will offer browser-based applications with scripts; some will have downloadable applications that you can execute; and in some environments, you will install the applications just like you would on a desktop PC.

What Defines a Good Application Environment?

The question arises of what a developer should look for when deciding what platform and application environment to design for. When we say *platform* here, we denote the device with its operating system (while the application environment contains the items within a device that are adjacent to the application). We illustrate the different parts of mobile devices in Figure 11.1. For mobile devices, there are some devices with closed platforms where no additional software can be added after production. Then, there are open platforms where third-party software can be downloaded and (sometimes) be installed. An open operating system can either be a foundation for an applications environment or it can be the applications environment itself. Examples of application environments are the operating systems EPOC, Windows CE, and Palm OS, but we can also include Java and WAP (both of which are independent from the OS).

First, let's look at some developers' wishes and then plunge into what the options will be in the future. Note that we are limiting this discussion mostly to applications for hand-held devices and not laptops and devices that can run a full-blown operating system such as Unix or Windows 2000/98/95.

Although there are many factors that determine which device and application environment to develop for, there are a few key features that a developer should analyze before making the choice:

Multithreading/multitasking. As we showed in Chapter 8, "Adapting for Wireless Challenges" it is beneficial to be able to run several tasks concurrently. Not only does it make coping with difficult network conditions easier, but it also gives a higher overall flexibility to the applications developer and user.

Low power consumption. Keeping the power consumption of mobile devices low is essential and is a very important, competitive aspect. The operating system and other parts of the application environment should be as power efficient as possible and make the majority of the battery power available to applications and communications hardware.

Communications integration. In order to get the most from the mobility and communications features of the mobile Internet, the application environment should facilitate access to the communications features of the device. This access includes items such as call control, *Quality of Service* (QoS), and choice of services.

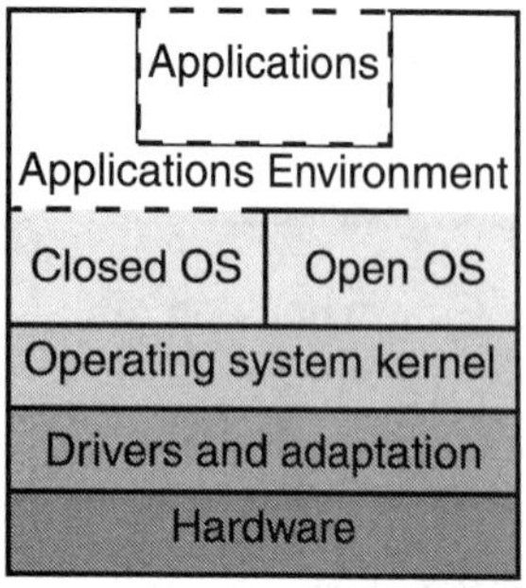

Figure 11.1 The different platform layers within a mobile device.

Stability and robustness. Mobile Internet users will probably not be as patient and forgiving of faults when executing applications as desktop PC users are. No one is used to rebooting the mobile phone just because the operating system crashes, and hopefully this problem will not emerge in the mobile Internet, either.

Broad range of devices. If the application is made for the mass market, the application environment should be, as well. The user is likely to stick with whatever is installed on the device at the time of purchase; therefore, the developer should make the chosen application environment supported by as many devices as possible. If the number of devices sold is great, then it is likely that the developer community also is large (which can be a great help with the development process). The easiest way of achieving this is by enabling portability of the applications, so that the same applications environment is used across many devices and applications can be written once to run on all of them.

We see a clear trend that the applications will become more advanced and also more diverse. The *third-generation* (3G) applications will span from the browser-based applications that we can see in *second-generation* (2G) networks as well to complex software implementations that run on an open operating system on the actual device.

Browser-Based versus Terminal-Based Applications

Technicians built the Internet around an architecture where the client (the Web browser) is a small (well, used to be at least) piece of software that works similarly on all platforms. A Macintosh user and a PC user can access the same Web site and even execute the same advanced Java scripts. The maintenance for the application developer is significantly simplified, because the developer makes all of the updates on the server side. This situation also eliminates the need for the costly production of CDs and distribution of boxes. With the limited functionality of the browser, it is also easier to control the spread of viruses and hacker attacks.

WAP introduces the browser-based applications to small mobile devices on wireless networks. WAP provides many of the features of the Internet in a format that is tailored to the smaller screens and limited input mechanisms. In addition, WAP enables the application to interface with the mobile network by offering access to call control and phone management features. The server then holds the majority of the functionality, and the application is said to be browser based or server based.

There are limitations to what you can do with a browser, however. You cannot store much on the device, and the developer does not have full control over the display. The ability to move graphics and control on-screen objects does not exist, which limits the user's experience. The application also requires that the browser be running, which limits the execution time to the periods of time when the user is actively using the device. With the advent of packet-switched networks that are always online, it would be nice to have an application that executes even when the user is not actively using the device. Imagine the device sending the pictures that you just took back to the office for publishing. The device could keep track of where you are and notify you when you pass by a city that is showing a great play. As devices become more advanced, the software developers who are used to developing advanced C and Java applications will want to apply those skills to mobile Internet devices, as well. Those applications might still have functionality on the server but also on the terminal. We therefore call these applications terminal-based applications. Note that the terminology is somewhat confusing here. A terminal-based application can actually be started by a choice in the browser. That link then downloads an application that executes on the terminal. The key is that terminal-based applications go outside the boundaries that the browser provides and run directly on the underlying platform. Table 11.1 shows some differences between the two types of applications.

For many developers, the question is, "How fat do you want your application to be?" The thin clients are easy to maintain (usually just a static browser), but they also have the limitations that we mentioned previously. Fat clients (terminal-based applications), on the other hand, give the developer full freedom, but the maintenance and distribution can be an issue. (Note that the words "fat" and "thin" are used here to indicate how much functionality and intelligence of the application that resides with the client.) In addition, not all devices will have open operating systems that enable application developers to run applications on them. Browsers, on the other hand, are on virtually every device. All major mobile phone manufacturers have announced that all of their devices will support WAP in the future, which opens a huge market. For those who want to reach the mass market, this choice is for you. A trend seems to be that applications are developed with one very thin client version (WAP) that is complemented with a terminal-based enhanced version for more powerful devices. Here, the content and presentation separation techniques of XML become handy, because you can access the same content with multiple client types by only altering the presentation templates. Even applications that host a client-resident part often have a server side that takes care of many of the heavier tasks, such as database management and analysis. Other clients that run on desktop or laptop PCs could then also access the same server-side application, as shown in Figure 11.2. Note that the mobile network in the figure might or

Table 11.1 Browser-Based Applications versus Terminal-Based Applications

BROWSER-BASED APPLICATIONS	TERMINAL-BASED APPLICATIONS
No installation needed	Could be installed or executed (when clicking a link) and then discarded
No distribution needed	Distribution of software needed; can use CDs or online channels
No or little execution of code on the device	Code can be executed on the device.
No direct access to the display or sound hardware	Limited or full access to different hardware parts of the device
Easy upgrade on the server side only	Updates can be difficult to distribute and inform users about.
Very limited virus risks	The virus risk has to be considered.
Operators like the application because they can control it if they install it.	Operators have little control over the client and might dislike the virus and business risks associated with it.
Has all of its intelligence and code on the server side	Has some intelligence and code on the client (and often, but not always, on the server as well)
No execution without a browser	The developer has full control over the application.

might not contain a service network, but the service network architecture is perfect for this kind of ubiquitous access.

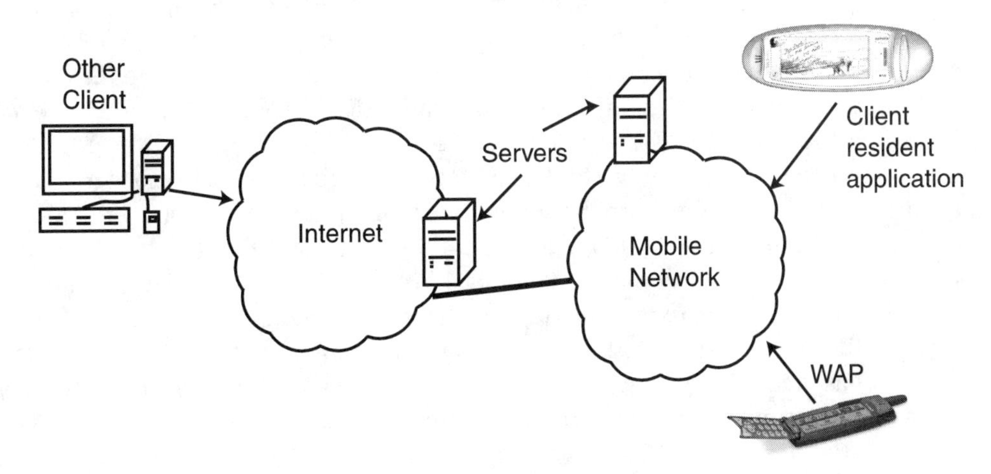

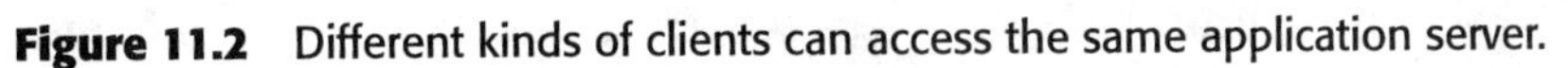

Figure 11.2 Different kinds of clients can access the same application server.

The Fight of the Operating Systems

Computer operating systems have always been a discussion topic in the *Information Technology* (IT) world, and everyone seems to have an opinion. The rivalry between MacOS, Microsoft Windows, and Linux users has sometimes led to heated discussions, and personal preferences seem as important as technical features. The media has always liked these fights and has enjoyed fueling the discussions. Now that the mobile devices are looking more and more like small computers, people are seeing the dawn of another debate about the mobile OS of the future. With heavyweights such as Microsoft involved, everyone expects a good fight. The question, however, is whether we really want a winner. Keeping a few really good contenders makes the competition harder and spurs innovation. If the applications are developed in an applications environment that is mostly independent of the underlying OS, such as WAP and Java, then things look very appealing.

In this book, we look at the three biggest operating systems as of late 2000 (measured by the number of supporting developers): EPOC, Windows CE/Pocket PC, and Palm OS. As a wild card, we have also added Pocket Linux, which is becoming an interesting alternative.

Palm OS

In the late 1990s, the Palm-sized organizer became a widespread productivity tool for many business people after a decade of struggle where predecessors such as the Apple Newton had failed. The Palm Pilot-series devices from Palm Computing (a wholly owned subsidiary of 3COM, Inc., at that time) led this development. The Palm organizers run an operating system called Palm OS, which developers created specifically for these devices. At the turn of the century, devices based on Palm OS had more than 80 percent of the *Personal Digital Assistant* (PDA) market. At the same time, licensees of the OS, such as Handspring and Sony, started to place volumes of devices on the market. Handspring added the Springboard module to their devices and opened opportunities for interesting plug-ins, such as MP3 players and Bluetooth communications modules. In addition, cradles that have *Cellular Digital Packet Data* (CDPD) modems appeared on the devices that did not have any expansion slots and on the Palm VII radio-enabled organizer. The Palm VII uses Mobitex, an old packet-based mobile system in a proprietary Palm.net service. Palm.net offers services from a number of contracted content providers that supply so-called *Palm Query Applications* (PQA). Each PQA uses *Hypertext Markup Language* (HTML) and a proprietary wireless access protocol called WebClipping. Palm in late 2000 also declared its commitment to the Palm OS as a wireless platform and announced the availability of Bluetooth-enabled devices in 2001.

The OS is designed to be small, fast, and intuitive to use, and it has the usage metaphor of an organizer. In other words, the devices were meant to be electronic versions of organizers, which prioritized easy access to tools such as Contacts, the To-Do list, and the Calendar. The efficiency of the OS enables Palm OS-powered devices to run several weeks without recharging the batteries.

Palm OS is a lightweight OS, and it can run on 16MHz processors with 2MB of memory. The OS is very fast, and the speed does not seem to be affected at all by the number of applications that are installed on a device. Although you enter Doze mode when nothing is happening (still keeps the screen lit) and you save battery life, the user never notices this feature. As soon as a user event takes place, the OS executes it—giving the impression that the application had the same high attention level all of the time. Those events are central in Palm OS, where everything circulates around an event loop. At any time during the execution, the OS is either executing a task that is derived from an event or waiting for the next event to occur (again, entering the power-saving Doze mode). While event-loop programming is commonly a part of programs in Java and C on other platforms as well, the novel thing is that here, everything is handled in this manner. The main programming language of Palm OS is C, but support for Java (we will provide more information about Java for mobile devices later in this chapter) was added as well in 2000. In addition, Palm OS lacks an ordinary file structure with directory trees. Instead, the data is stored in records and databases, which you can access quickly. Again, these concepts are not new to experienced programmers, but you can choose other ways to perform tasks on many other platforms. A problem with the event-based architecture used (at least, up to Palm OS 3.5) is the lack of multitasking. There is no way that Palm OS can execute two commands at the same time. As we discussed in Chapter 8, "Adapting for Wireless Challenges" this situation puts some serious limitations on the flexibility of communications. Creating an application that fetches data in the background is cumbersome, for instance, and interruptions will block the entire application. Some of this process is compensated by extreme user control, where it is almost always possible for the user to intervene.

The initial Palm OS-based devices communicated almost exclusively via synchronization with a PC. This synchronization included not only the Calendar, the To-Do list, and other personal information, but it also included downloading of a number of favorite Internet sites via the AvantGo browser. AvantGo started as an experiment by Silicon Valley programmer Linus Upson, who wanted to have the New York Times crossword on his Palm device. After managing to get the synchronization to include the New York Times site, Linus and some of his friends generalized this concept to download Internet pages to Palm devices (this process has been generalized to support Windows CE as well) to be accessed even when offline. Although Palm devices are migrating toward being more and more connected, the AvantGo concept of facilitating connectivity for casually connected devices is very appealing. As we saw in Chapter 8, "Adapting for Wireless Challenges" even in the 3G networks there will be times where the data cannot get through to the user, and it is then convenient to still have limited application functionalities on the device.

The Palm serial communications port enables the Palm to talk with external devices, such as bar-code readers and communication cradles that support CDPD, GPRS, and Bluetooth. This functionality opens Palm OS-powered devices to entering the mobile Internet, either as integrated concepts (for example, a GPRS cradle) or as divided concepts (for example, by using Bluetooth). In addition, there are two *Application Programming Interfaces* (APIs) available for network communications: the Net Library and the Internet Library. The Net Library supports TCP/IP and UDP/IP and enables the application to set up regular sockets. Some developers complain that the Palm OS communications stack is too rigid, because there are few parameters that can be configured (for instance, in TCP). We have seen innovative solutions to this problem where additional transport layer functionalities are added in order to give full control of buffers and timeouts.

Palm OS's market share alone makes it a strong contender in the race for mobile OS dominance. The key question is whether you can scale the OS to support the multimedia and flexible communications requirements of 3G networks in order to keep the lead in the race. Will the appraised organizer-user metaphor prevent the OS from penetrating other segments?

Windows CE

While Palm OS was designed with a certain usage scenario (the organizer) in mind, Windows CE was designed by Microsoft to be a mobile version of its popular desktop OS. Although Windows CE is not a scaled-down version of Windows NT or 2000/98/95, all of those operating systems still have many similarities. Windows CE was indeed developed from scratch in order to fit mobile devices while still keeping most of the look and feel of its bigger desktop brothers. The result was an OS that did not have all of the features of a desktop OS and that still lacked the robustness and speed that was needed for a mobile OS. Consequently, Windows CE remained a niche player in the hand-held space. After having been a small niche player in the hand-held space, Microsoft put in massive artillery in the late '90s in order to ensure that the company would become a major force in the mobile OS field as well. This situation culminated in the release of Windows CE 3.0, which powers Pocket PC devices. This release held many major improvements in just about every part of the OS, and people started to see Microsoft catching up with Palm. Pocket PCs are powerful multimedia devices with color graphics and stereo sound support built in. Typical processors run at hundreds of megahertz, which makes it possible to encode and decode many multimedia formats.

Although Windows CE has had a hard time in the consumer segment, vertical appliances such as TV set top boxes have been really successful. Three versions of Windows CE exist: one for vertical applications such as cars, one for hand-held PCs, and the Pocket PC version. The discussion in this book is limited to the Pocket PC version.

Although the rich multimedia support and powerful hardware are important strengths of Windows CE, the strongest part is its development environment. Windows CE uses Win32 APIs, and anyone who has been developing for Microsoft's other platforms will feel at home with Windows CE as well. Some Windows 95/98/2000/NT programs can actually be recompiled for Windows CE and work, although that is something that I definitely do not recommend as a habit. As always with mobile devices and operating systems, the display and memory properties are significantly different.

The use of multithreading/multitasking significantly increases the possibilities for developers to create good applications. Communication with networked devices can be located on separated threads in order to isolate the rest of the application from disturbances, such as interruptions. You can use one thread to stream stock quotes from a server while another one updates a graph on screen and lets the user manipulate it.

Most Pocket PCs have shorter battery life than the competition (in the range of days or even hours when intensive applications such as streaming media are used). The APIs do support power management functionality, but the color screens and powerful CPUs drain lots and lots of power from the battery. As more and more devices become wirelessly enabled, it will be interesting to see how long the battery life will last. Maybe it is all these advanced features rather than the OS that limits the battery life? A good solution is probably to use Bluetooth chips in Pocket PCs and let a Bluetooth-enabled GPRS/3G phone take care of the communication with the network (the divided concept). Many Windows CE devices have a compact flash or PCMCIA slot that you can also use for external peripherals, such as wireless modems and *Global Positioning System* (GPS) receivers.

The communications features of Windows CE are intuitive for those who are familiar with the Win32 communication APIs. Most of those APIs have been ported directly to the smaller platform, however, and it is likely that the same code that works for the desktop operating systems will work on a Pocket PC without modifications. This feature gives developers access to a huge library of written code, either through old Win32 projects in house or on the Web. Many of the lessons that we learn on the software level can then be reused for the mobile world, but you should take special care and pay attention to the advice in Chapter 8, "Adapting for Wireless Challenges."

The drawbacks with Windows CE have so far included high battery consumption and a lack of multimedia demand and user friendliness (sometimes requiring the user to go through many clicks in order to start an application). With the advent of 3G networks, where multimedia applications play a central role, maybe the time has come for Windows CE.

EPOC

While both Palm OS and Windows CE were primarily designed for hand-held computers and PDAs primarily, EPOC has been designed as a communications-centric OS for mobile devices. The U.K.-based company PSION migrated in the early 1980s from designing software for computers such as the classic Sinclair ZX Spectrum to making hand-held computers. The first device was launched in 1984 with some simple applications such as a database, diary, alarm clock, and a simple programming language called the *Organiser Programming Language* (OPL). OPL was a Basic-like language that borrowed the name from the first device, the Organiser. The device evolved during the 1980s until the next major step was taken in 1991 when the PSION 3 organizer was launched. Now, developers could develop in legacy-language OPL as well as in more widespread alternatives such as Assembler and C. This situation led to a more widespread developer support, and many interesting applications surfaced. The PSION Series 3 was a good mobile platform with low power consumption and high stability. The input mechanism was limited to a small keyboard, but navigation and control were still easy through intuitive design.

In moving from 16 bits to 32 bits, PSION started the development of a new platform in 1994. With the advent of the Nokia Communicator (which used another operating system from Geoworks) around 1997, PSION saw the possibilities of the combined hand-held computer and mobile phone. The initial plans of purchasing a *Global System for Mobile* communications (GSM) company in order to get into that space were abandoned for the alternative of a joint venture with the three biggest mobile phone manufacturers in the world: Ericsson, Motorola, and Nokia. The Symbian joint venture was announced on June 24, 1998, and was joined by Panasonic (in other words, Matsushita) shortly afterward. The aim was to create a communications-centric OS, EPOC, out of the 32-bit core that PSION had developed. The first EPOC-enabled products were released in mid-1999: the PSION 5mx and the Ericsson MC218.

EPOC targets smartphones and communicators, where there is close interaction between the communications equipment and the software. Because even these two categories span many devices, Symbian has created a number of reference designs (that developers can target. At this writing, three such reference designs exist: Pearl, Crystal, and Quartz.

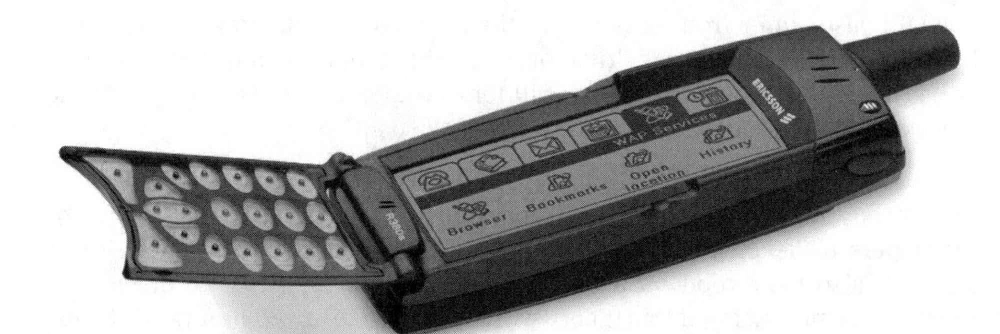

Figure 11.3 The R380 smartphone.

Pearl is a smartphone where the display is horizontal (the actual resolution differs between devices). These devices will still look and feel a lot like phones, but they also have extended PDA and applications functionalities. Figure 11.3 shows roughly how Pearl devices will look, although the R380 is *not* developed according to the reference design.

Crystal is the data-centric design with a similar form-factor as the original Psion 5. These devices enable applications and data and access to the mobile Internet but also have voice capabilities.

Quartz fits right in the middle between the phone-centric Pearl and the data-centric Crystal. These devices are commonly called communicators and put an equal weight on voice and data capabilities. The concept communicator in Figure 11.4 shows a Quartz communicator with a color screen, GPS receiver, and built-in GPRS and Bluetooth.

Today, development for EPOC mostly takes place in C++, but Symbian is starting to increasingly push Java, as well. For EPOC 5.0, the Java version is built on Java 1.1.4 (Personal Java), and the Symbian developer site (www.epocworld.com) has some good technical papers on the topic. You can still use OPL, but developers who are familiar with C++ have no reason to migrate to OPL. Both Java

Figure 11.4 Quartz communicator concept device.

and OPL also suffer from difficulties when accessing some lower-level EPOC features, but with the strong drive for Java throughout the industry, this situation is likely to change. EPOC was built for development by using C++, and the most widely used tool seems to be Microsoft Visual C++.

Much like Windows CE, EPOC supports multimedia and multitasking and often runs on powerful processors. Its reference design system makes it possible for developers to develop one application for a Motorola Quartz communicator that will also look good on a Panasonic Quartz device (if these companies release such products). If this theory works in practice, it will be a valuable feature that makes the design of good-looking applications easier.

Because EPOC was designed at the outset to be a good OS for communicators, there is plenty of documentation and support available for that part. There is even a dedicated *software development kit* (SDK) called EPOC Connect, which enables convenient development in many languages. EPOC was the first OS with a WAP browser built in (MC218 in 1999), and many developers who start working with EPOC do it for the communication parts and for the support of the wireless giants.

The success of EPOC depends largely on the interest from device manufacturers and applications developers. The latter are, of course, very interested in making sure that devices will be available. EPOC seems to fit between the Palm OS and Windows CE in many aspects and as devices start to appear on a broader scale, it will be a strong contender in this race.

Pocket Linux

While the giants Palm, Microsoft, and Symbian fight for the throne of the mobile OS, others are looking at other alternatives. Linux has long been popular in the IT world for its open source approach, but the widespread adoption of the OS has yet to be seen. With the tightening competition in the mobile Internet device space, the thought of an OS that you can use for free is very appealing. Mobile devices also generally have a lower manufacturing cost than a full-size PC, which means that cutting costs for individual parts is more important.

There are a number of interesting projects that aim for bringing Linux to handheld devices and maintaining it. An interesting one is Pocket Linux, which was shown running on a Compaq Ipaq during summer 2000. The kernel has been re-engineered for small devices and optimized in many ways. Applications can be written in Java, because it uses the Kaffe open source Java implementation (which hopefully makes it easy to run existing Java applications on it). In addition, everything in Pocket Linux is based on XML—not only the applications, but also things such as the system databases and e-mail.

In January 1999, a Silicon Valley-based company called Transmeta launched a new and revolutionary processor called Crusoe. Crusoe is extremely power efficient and still enables power applications to run on it. In addition, it can execute any x86 application by a software translation of instruction called morphing. With Linux creator Linus Torvalds as a part of Transmeta, we foresee Linux to be installed on many devices that run on Crusoe.

Linux is an interesting contender, and it remains to be seen if that is what it will remain to be. Its breakthrough and widespread adoption depends mostly on the support of device manufacturers. If users have to install Linux themselves and throw out the existing OS, then it will remain a marginal player. Linux and Transmeta also illustrate how we might be overestimating the importance of the OS. Maybe Java and WAP will make the mobile OS irrelevant for applications developers.

Who Needs an Operating System?

This question is becoming highly relevant as we see the migration to a world of standards and platform independence. No one will develop an application that can run on GPRS only; rather, developers will design the application to run on all packet-based mobile Internet networks. You can access WAP and other XML applications on Windows CE devices, and people who use Palm and EPOC can also have this access. The standardized APIs in the service network use CORBA and enable many different platforms and languages to interact. The AT commands that enable applications to talk to the *Mobile Terminal* (MT) can be used on any platform. Will the applications developer in the future even care what the underlying operating system is? From the software developer's point of view, it would be best if the application could be written for one device and one network. Then, you would only need to perform minor tweaking in order to make it work on other networks.

While this goal might remain a dream, many efforts are being made to consolidate the application environments and to standardize as much as possible. Since the mid-1990s, one of the efforts that has been made is Java, which removes the importance of the operating system.

Java for Mobile Devices

Java promised platform independence when it was launched, but this goal has been hard to achieve so far. With the modifications that are needed to adapt to different screen resolutions and user interfaces, developing in C and isolating the graphics and hardware dependence was almost equally efficient. Java also quickly gained the reputation of being big and fat. I wrote my first Java program in 1996 and produced a "Hello world" implementation that boasted an impressive file in the megabyte range! I then learned to configure the compiler better, but Java still traditionally requires a lot from the underlying device. Therefore, many people doubted that it was possible when Sun Microsystems announced Java 2 Micro Edition for mobile devices in June 1999. How could anyone even dream of squeezing the Java monster into a mobile phone?

With the advent of Java 2, Java split into three editions that were targeted at different devices and applications:

Java 2 Enterprise Edition (J2EE) targets large enterprise applications and servers. The use of CORBA makes it fit into the service network rather than inside the device. We do not describe J2EE in detail in this book.

Java 2 Standard Edition (J2SE) is a natural evolution of Java 1.x, which provides an easy migration for existing Java developers. The size is still too large for most mobile devices, and we will not describe J2SE in detail in this book with the exception of Personal Java. Personal Java is targeted at communicators and powerful PDAs (for instance, the EPOC Quartz reference design).

Java 2 Micro Edition (J2ME) is the scaled-down Java version that is suitable for mobile devices.

Figure 11.5 illustrates the different Java 2 editions.

This figure illustrates how the basic J2ME is extended with two configurations. J2ME (at this writing) consists of two configurations: *Connected Device Configuration* (CDC) and *Connected Limited Device Configurations* (CDLC). Configurations define a minimum platform for a group of target devices, each with similar properties in terms of memory, size, and processing power. In other words, a configuration consists of the classes and virtual machine features that are targeted toward a certain group of devices.

CDLC is the configuration that is most appropriate for small, mobile Internet devices. Considering the specific needs of this group of target devices, the configuration is very compact and limited. CDLC is defined as a subset of CDC, as shown in Figure 11.6.

CDC is targeted at somewhat bigger devices, such as set-top boxes and in-car systems. All of these devices have more generous environments in terms of processing power and battery power (many of them do not have batteries but instead use nonlimited power sources). The other main difference between CDC and CDLC is that CDC uses the standard *Java Virtual Machine* (JVM) while CDLC uses a smaller virtual machine, the *K Virtual*

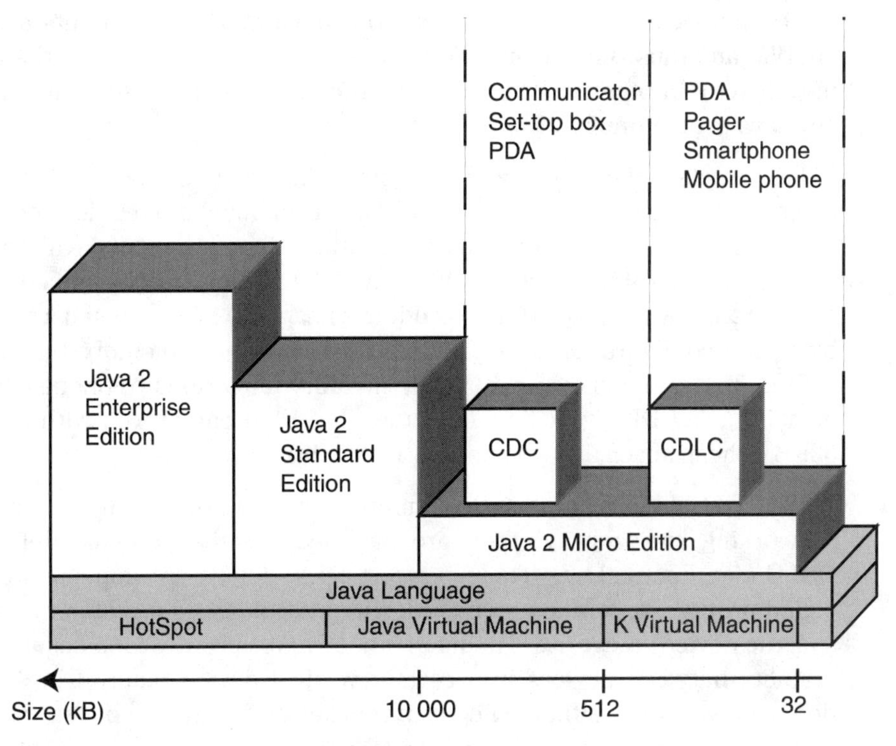

Figure 11.5 Java 2 editions and configurations.

Machine (KVM). K means that the resulting footprint on devices should be in the range of kilobytes rather than in megabytes.

You should interpret Figure 11.6 as J2ME (in other words, CDC and CDLC)—a Java edition that has been adapted to the small devices. This adaption mostly took place by removing things that were not considered needed or feasible for the small devices. In doing so, the need for some specialized classes has arisen (and for those that are not part of the larger Java editions). In this section, we will look more closely at CDLC, because this point is where most of the work

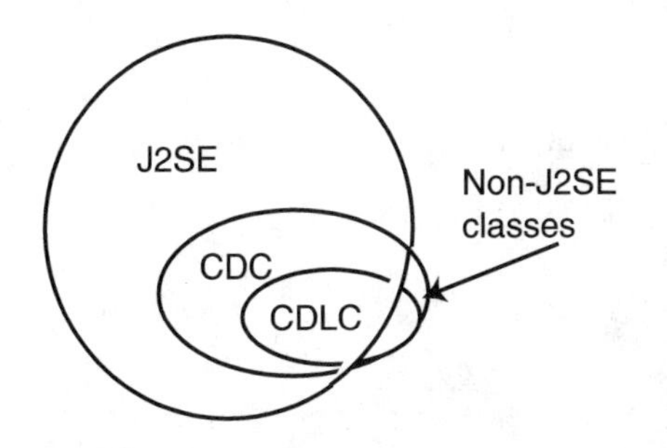

Figure 11.6 Not all CDC and CDLC classes are included in other Java editions.

has been done in order to fit Java into tiny devices. CDC is built upon the regular JVM and thus runs on similar devices as Java 1.x (memory in the range of megabytes). In other words, developing for CDC is pretty straightforward for any Java programmer.

CDLC targets devices such as PDAs, smartphones, and mobile phones, which can spare at least 128KB of memory. The virtual machine that is used, KVM, is extremely compact when compared to other incarnations of Java. The actual footprint of KVM is between 40KB and 80KB (depending on compilation options and the target platform). Adding a heap size of 128KB and a few tens of bytes for configuration and class libraries, a typical memory budget totals 256KB. This value is even within the memory limitations of phone-centric 3G devices. The CDLC specifications state that KVM can be run with as little as 32KB of heap space, but you usually need a bit more.

While CDLC defines some basic common denominators (such as input/output, networking, and security), there are many features that you must define above CDLC (see Figure 11.7). These more specific details are implemented in so-called profiles. A profile extends a configuration and addresses the needs of specific device categories. The main objective for profiles is to ensure interoperability between devices in a certain vertical device family. The idea is to define classes that fulfill the needs of this device family without sacrificing the code's portability.

The *Mobile Information Device Profile* (MIDP) is the first profile to be defined. The target device group includes mobile phones, pagers, and PDAs. The characteristics of an MIDP device are the following:

- A monochrome or color display that is at least 96 pixels wide and 54 pixels high
- Input mechanisms consisting of a touch screen, a keypad, or a conventional keyboard

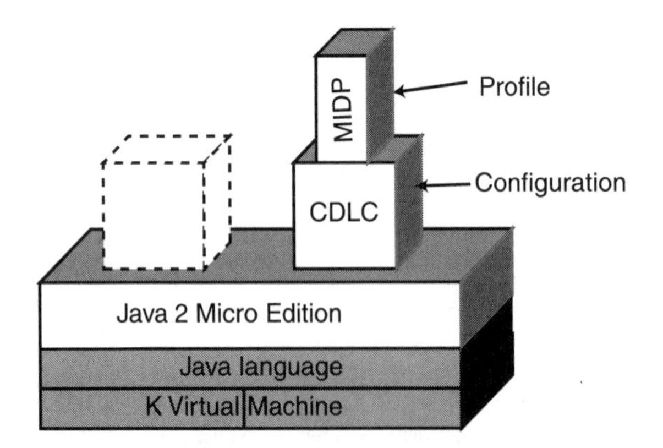

Figure 11.7 A profile sits on top of the configuration.

- A wireless network connection that has limited bandwidth (the connection does not have to be always online and thus can include 2G networks)
- 128K of memory that remains even when the device is turned off (nonvolatile) in order to store the MIDP components, 8K of nonvolatile memory in order to store persistent data, and 32K of volatile memory for the Java run time

The key contribution from MIDP is the definition of user interface handling. This handling includes most aspects of the items that are displayed on the screen, including text, graphics, and user interaction. Other parts of MIDP include extensions of the CDLC network-handling features (mostly for HTTP) and nonvolatile storage records.

Various terminal vendors (for example, Motorola) and operators (such as NTT DoCoMo) have announced strong support for KVM and Java applications for mobile devices. All issues are not yet solved, however, and as of late 2000 it is impossible to use the supplied SDKs to make applications that run across all devices. During a transition period, it is likely that the user interface code will have to be adjusted for the desired platforms.

Hopefully, profiles such as MIDP might be capable of solving this problem in the future and will enable the dream of write once, run on multiple devices to come true.

Terminal Capabilities and MExE

In order to standardize components for mobile devices, 3GPP developed the *Mobile Execution Environment* (MExE) specification (3GPP TS 22.057 and TS 23.057) that defines a framework for such capabilities. The aim is for the applications developer to be able to develop for a certain class of devices with a common classmark, rather than for specific hardware. This ambition is sparked by the advent of technologies such as WAP and Java, both agreed by 3GPP to be fundamental for future mobile devices. Today, developers check what WAP version a set of devices supports and develop for it, but what happens when Java and possibly other terminal capabilities arrive? 3GPP wants to avoid a situation where there are so many permutations of mobile device capabilities that it would be impossible to develop applications for more than one device at a time.

In the first release, as part of the 3GPP release 1999 (the first UMTS release), WAP and Java are supported. As other groups standardize these technologies, the majority of the MExE work has been to choose which versions of the two to support and to add security features. These features include security certificates, which indicate that an application is secure and also where it can be executed. In addition, two classmarks are defined:

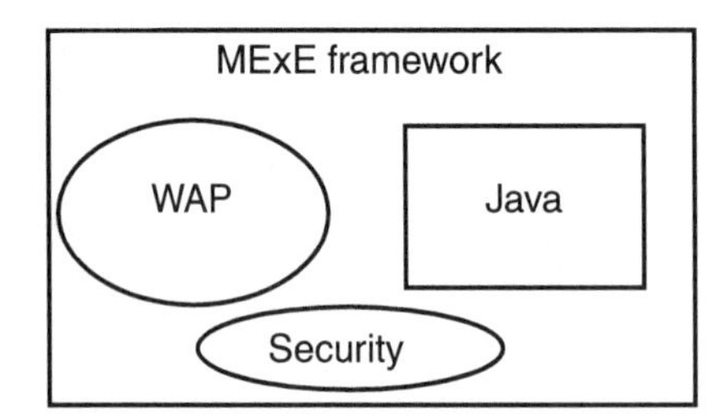

Figure 11.8 MExE includes WAP and Java.

Classmark 1 defines devices that support WAP 1.1 and newer versions that are backward compatible (in other words, all WAP versions that have been defined to date).

Classmark 2 defines Java-enabled devices. This Java version is Personal Java, however, which is based upon Java 1.1.4. This issue is significant, because Personal Java puts heavy system requirements on the device (for instance, memory in the range of megabytes). Future MExE versions (see below) will support kJava and other features, shown in Figure 11.8, where MExE is depicted as a framework that includes several technologies.

As an example, an application can be defined as a MExE Classmark 1 application. In other words, devices need to support Classmark 1 to be capable of running the application. Another application might be defined as a MExE Classmark 1 and Classmark 2 application and consequently should only run on MExE Classmark 1 and Classmark 2 devices. Thus, the different classmarks are not subsets of each other but are instead distinct mobile device capabilities.

Perhaps the most interesting feature of MExE is added in the second release of the standard (3GPP release '00), where Classmark 3 is introduced. Classmark 3 defines the support for kJava according to the description in the previous section. Even though this is a later release of the standard than the first WCDMA version, it is expected that many 3G handsets, even some of the first ones, will support Java through MExE Classmark 3. This is because it is much easier to take a standard like MExE from the standard into a product than things like radio interfaces.

Although MExE is a standard in the works and not all parts of its vision have come true yet, it will be a very interesting part of application developers' mobile Internet toolbox.

Summary

The first thing that comes to mind when talking about application environments is the operating system of the device. Developers have become accustomed to the desktop PC world, where the operating system decides the development characteristics. For mobile devices, there are some devices with closed platforms where no additional software can be added after production. Then, there are open platforms where third-party software can be downloaded and installed. Java 2 Micro Edition and Personal Java have emerged as compelling application environments that would make the application more or less independent of the operating system.

CHAPTER 12

Security

Now that we have a greater understanding of the wireless networks and their accompanying enablers, we are closer to filling our toolbox of wireless application tools. These tools will enable us to create the applications that we so dearly need for the success of the mobile Internet. For some of these applications, achieving sufficient security is just something that we do automatically. An application that monitors the weather for those destinations that are on your next itinerary might not be security sensitive, and other things might be of higher priority. On the other hand, if you are developing a mobile banking application, achieving sufficient security is crucial even for the entire feasibility of the project. Security is very large and important, and this chapter aims to explain some of the fundamentals in order to aid developers with making the right decisions.

How Secure Does It Have to Be?

When we ask users this question, most will answer that it has to be totally secure in order for them to use the site and to trust it. Just the thought that there is a tiny chance that someone could steal from us makes us terrified. This situation is especially true with new technologies, where we feel that there are so many people who know much more than the rest of us and who are just waiting for the right time for fraud. The Internet has really given life to the legends of hackers who can break into pretty much any computer system and change or even destroy the content. Sometimes, however, it feels that we are much too paranoid about the new technologies and trust the old ones too much. One good example is the use of credit cards. Most people are very careless with credit card slips in bars and restaurants but are scared about leaving their credit card number on the biggest online retail site over a secure link. I do not know anyone who can swear that he or she has never forgotten a credit card slip somewhere, and any one could then pick it up and use it pretty easily (both the credit card number and the expiration date). This method is ironically also the easiest way of fraud on e-commerce sites: finding someone's credit card slip and shopping online with it. In addition, the servers themselves, on which many e-commerce sites keep the credit card numbers, are also sometimes insecure. Thus, it is much easier to hack a server and steal credit card numbers, than to eavesdrop on SSL connections. The interesting part here is that the Internet in itself in this case is not the problem, and the eavesdropping of traffic is much more cumbersome, but still the blame for the fraud is on the network in lots of media.

Having lived in both the United States and Europe during the past couple of years, I have found a tremendous difference in trust of the new technologies in the two continents. In the United States, people have been leaving their credit card numbers over the phone for years, and the Internet is no different. In Europe, on the other hand, many people stay away from online shopping just because they fear fraud. In those cases where people take the step and start using e-commerce sites, the goods are often shipped to a nearby post office where the actual monetary transaction takes place. Similarly, the online bank that I use in the United States only requires a password in the Web browser, while some Swedish banks will give you a small code box that you can use to access the online service. Perhaps this reason is why e-commerce has penetrated society more easily in the United States than in Europe, and this example shows how important the user's perception is with security issues. Consequently, users have to be very comfortable with applications developers and content providers in order to facilitate successful e-commerce and *mobile commerce* (m-commerce) application implementation.

We should also add that in the majority of credit card fraud cases, the user is not hurt economically. Rather, the credit card company takes the fall. Consequently, the credit card companies are some of the major drivers behind security on the Web (and now also the mobile Internet).

One of the more common mistakes that you can make when implementing security solutions is suboptimizing one part and neglecting another. In this chapter, we will describe three parts of mobile security that all need to work together satisfactorily in order to make the overall application secure. For instance, it is worthless to have a powerful encryption algorithm if users still choose their name backwards as their password. These kinds of security flaws are actually some of the most dangerous, because they give everyone a false feeling of security.

While the technical solutions are sometimes tricky, it all comes down to the user's perception in the end. Making the user comfortable with the level of security and with the application as a whole is the trickiest task. Perception and trust are everything.

Securing the Transmission

Although the previous section implies that people tend to over-dramatize security risks, there are severe issues surrounding today's Internet—and malicious users can tap into systems and eavesdrop or modify traffic. We need powerful security mechanisms for communications technologies (here, the wireless networks) in order to ensure that higher layers can rely on secure transmission properties of the networks.

Maintaining the integrity of a message (ensuring that it is received as it was sent) and preventing eavesdropping are the main tasks of cryptography. In brief, we perform confidentiality and eavesdropping prevention by utilizing encryption, while we keep the integrity intact by using checksums, authentication codes, and digital signatures.

Authentication

Authentication aims to verify the identity of communicating parties. In a wireless network, this procedure is commonly done both at the network layer and by higher-layer protocols that the application uses. We can perform authentication with either a public key or a secret key. An example of a public key is an Internet banking application, where the user downloads a certificate that the bank offers as proof of its identity. The user then presents this certificate when connecting to the banking service, and he or she can only use that certificate on that bank's Web site. The use of public keys creates a need for a *Public Key Infrastructure* (PKI) in which a key can be associated with an identity (a user, a company, a machine, and so on) and issued by a trusted party. A certificate commonly includes both the identity of the party (name and unique ID) and the public key. A big and well-known *Certificate Authority* (CA), such as VeriSign, usually issues the certificate and ensures its correctness. An application developer can turn to a CA to buy a certificate for the application in order to make users feel comfortable with the service. Usually, the certificates are only valid for a limited time to make control easier.

Encryption

A common way of protecting information from eavesdropping is the use of encryption, where the message is coded in a way that only the sender and

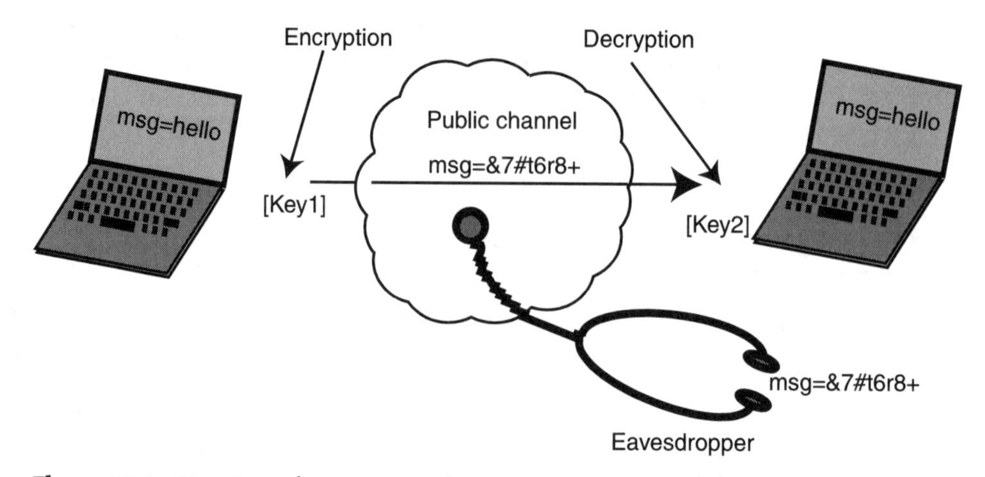

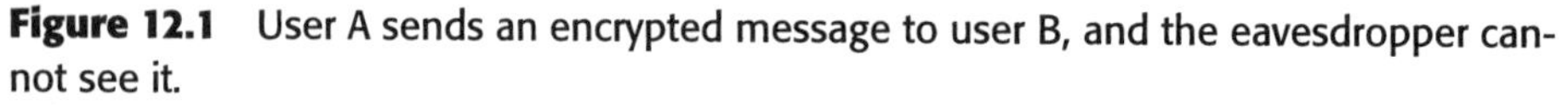
Figure 12.1 User A sends an encrypted message to user B, and the eavesdropper cannot see it.

the receiver can access the data. We illustrate the basics of encryption in Figure 12.1.

User A sends a message (*msg*) over a public channel over which user M (the malicious user)could potentially eavesdrop. The message is intended for user B and is encrypted in order to ensure confidentiality. The sender (user A) encrypts the message using *key1* in a ciphertext before sending it, which makes the message unreadable to M, who tries to read it. User B then uses *key2* to decrypt the ciphertext and can then access the message that user A sent.

If *key1* and *key2* in this example are equal, the system is said to be symmetric, while systems that have different keys are called asymmetric. One case of asymmetric systems is the instance where *key1* is made public (*key2* must always be secret, of course). A prerequisite is, of course, that it is impossible to derive *key2* from the public *key1*. The problem then arises of how to distribute the keys to the receiving parties, and we describe this situation in the next section.

While encryption ensures that no one can listen to the traffic and extract confidential information, it does not protect against adding or removing information. Someone who manages to listen in on the communication in this example could potentially change the bits in the message and thereby also change the content of the information. A remedy is to protect the message integrity.

Protecting the Message Integrity

Checksums and *Message Authentication Code* (MAC) fields (not to be confused with the Medium Access Control protocol) usually keep message integrity. The MAC field works like a regular checksum, where bits are added at the end of the message by applying an algorithm to the message. The recipient then uses the same algorithm in order to ensure that the message has not been altered. If someone changes some bits in the message, the MAC field will not match the rest of the message, and the recipient will know that something is wrong and will discard the message (see Figure 12.2).

The MAC field security mechanism is a symmetric technique, because the sender and the recipient use the same MAC field in the process. An example of asymmetric coding is the use of digital signatures that employs the private secret key to sign the message. A signature is known to the public (could be downloaded from a Web/WAP site) and can be accessed and verified by anyone who has an authentic copy of the corresponding public key.

GSM/GPRS/3G Network Security

In GSM and its evolutionary systems, the operator controls all of the security keys and authentication methods. The *Subscriber Identity Mechanism* (SIM) card holds all of the vital subscriber information and the keys that we need. All of the dedicated traffic (voice and data) is encrypted and securely protected. The authentication calculations are only calculated in two places: by the SIM card and in the *Authentication Center* (AUC), which means that the operator

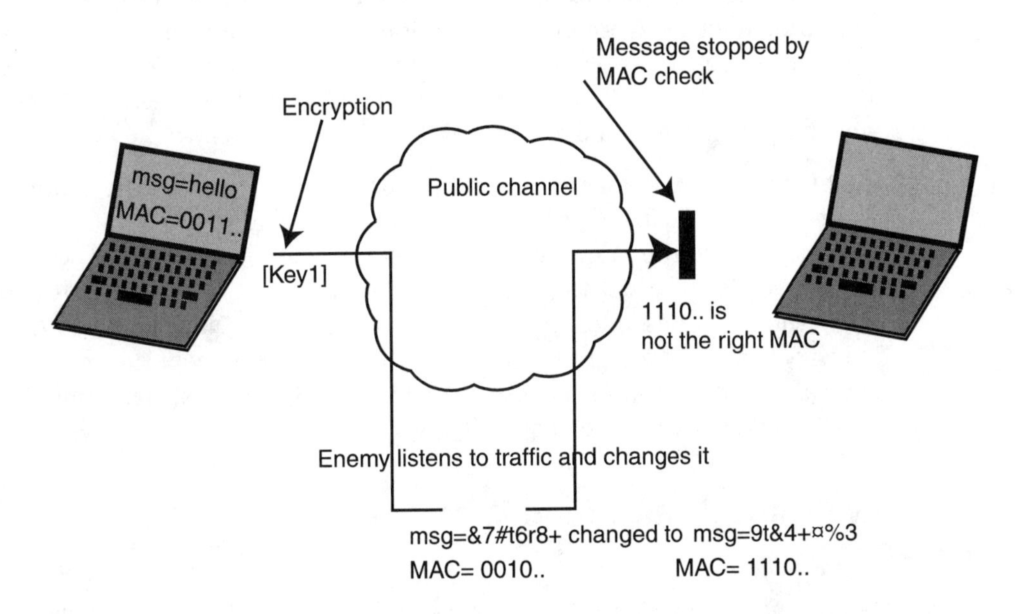

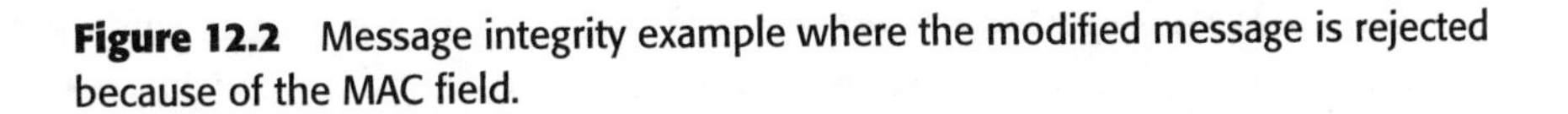

Figure 12.2 Message integrity example where the modified message is rejected because of the MAC field.

has full control over the security (the operator gives the subscriber the SIM card when signing the subscription). As a result, the wireless link of these systems provides one of the most secure transmission mechanisms available.

Enabling Security in Higher Layers

The basics of security are based in the network layers, where the actual information is sent. In addition, there are usually firewalls deployed that ensure that only the desired traffic passes.

As we saw in the previous section, there are good ways to ensure the integrity and confidentiality of traffic on the networks, but we need the support of higher layers to make it possible. If we want to use keys for more than just the wireless part of the transmission—and if we want to authenticate more than just the device and the subscription—then we need protocols to facilitate this process.

Algorithm Decision

At the start of a communication session or when initiating a secure application, the communicating parties need to decide on a common set of algorithms to use throughout the session. This decision includes the encryption algorithms and data integrity protection as well as how to exchange keys. Exchanging keys is especially sensitive when symmetric keys are used, because finding out what a key is means finding out what the other key is, as well.

If malicious users knew the algorithms that we were using, then it would not matter much if we had a powerful encryption. One might want to use the algorithm to decode the information, but it is not quite that easy. The key would also be needed in order to pull it off. Earlier it was common to try to hide as many things as possible in order to ensure security. Now it seems that the industry is moving toward keeping everything known except the key. Preventing leakage of algorithms is very difficult and relying on keeping them secret could create a false sense of security.

Security Protocols and Their Wireless Usage

Developers are most likely to get involved with the following protocols when working with wireless applications.

Wireless Transport Layer Security (WTLS). WTLS is part of the WAP stack and enables the use of certificates, as we described in Chapter 7, "The Wireless Application Protocol (WAP)." WTLS is an enhanced version of TLS

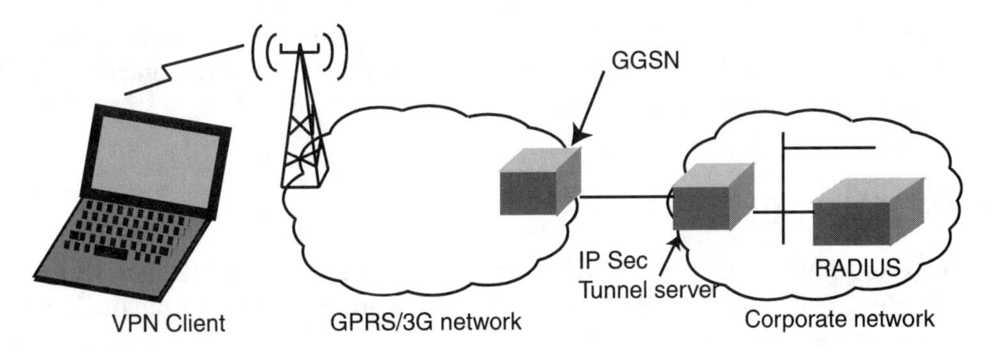

Figure 12.3 GPRS VPN access.

(although they are largely the same), formerly known as the *Secure Sockets Layer* (SSL), which might be used by some devices that lack WAP support (mainly more powerful devices such as laptop computers).

IPsec. IPsec enables IP layer security for a variety of bearers, including connectionless ones. IPsec includes encryption and other cryptography features. The standardization of IPsec was finalized in 1999, and deployment then started. *General Packet Radio Services* (GPRS), for instance, includes support for IPsec, which is especially useful when the backbone is shared between several operators and when you are using *Virtual Private Network* (VPN) applications. A VPN application enables corporate users to access internal information even when it is not within the physical location of the company. People generally view this functionality as one of the more obvious successful applications, where GPRS-enabled laptops (via a PC card or a Bluetooth-enabled GPRS phone) are always connected to the corporate intranet and e-mail systems (see Figure 12.3).

In Figure 12.3, the GPRS-enabled laptop connects directly to the corporate intranet, which assigns an IP address. The RADIUS server authenticates the user and gives him or her access to corporate resources. You can also use other solutions where DHCP is used for IP address assignment and RADIUS only authenticates the user.

In addition, there is a wide range of proprietary security solutions that solve specific problems.

Security Issues

Although mobile Internet networks are generally much more secure than their wired counterparts, there are still some issues that you need to consider. Most of these issues are fairly easy to handle once they are known and considered. The first one involves the protocol translation in the WAP gateway, and the second involves Bluetooth's lack of user-level security.

WAP Security Issues

If we recall the description of the *Wireless Access Protocol* (WAP) gateway in Chapter 7, we remember the way that protocols are converted within the WAP gateway (as seen in Figure 12.4).

The WAP gateway converts the TCP/IP protocols into WAP protocols, which includes a translation of the security features. We use Transport Layer Security (TLS) between the WAP gateway and the content server, and Wireless Transport Layer Security (WTLS) between the WAP gateway and the WAP mobile device. Right when we make this conversion, we even have to encrypt data for a brief period of time during the translation. People have voiced their concerns over this security procedure, because someone could potentially hack the WAP gateway and gain access to this information. WAP gateway manufacturers have been very active with designing the gateway in such ways to minimize this risk. These efforts include doing both decryption and encryption in the same

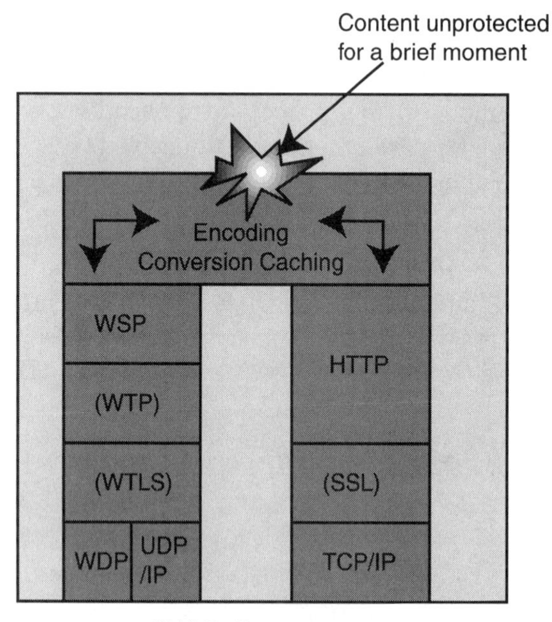

Figure 12.4 WAP gateway anatomy with a security flaw.

process internally, thus not storing the unprotected data in persistent memory and minimizing the time that the data is unprotected. Therefore, this issue comes down to whether the application developer trusts the mobile operator (or whoever owns the WAP gateway) and how much control the user wants to have over it.

Despite the precautions and trust issues of mobile operators, some applications developers who use applications that have high security requirements are solving this problem by hosting the application themselves. Examples include mobile banking, mobile brokerage accounts, and *mobile commerce* (m-commerce) sites. Those who decide to take this step need to be aware of some of the consequences, however:

- You need to work on maintaining the WAP gateway. This node is not a simple PC that any *information technology* (IT) personnel know how to configure and run.
- Some WAP handsets only enable the configuration of a single WAP gateway. In other words, the handset needs to be reconfigured when changing to your application and its WAP gateway. Some handsets enable users to define profiles that they can switch between. You must weigh in this added complexity.
- The WAP gateway costs money and probably needs to be upgraded when new releases of the standard arrive.

The advantages include the security options that we discussed previously but also full control of all other features of the gateway. The application owner does not have to rely on great availability from the operator but can be in 100 percent control himself or herself. As WAP usage increases, the private gateway is only affected by the traffic of this dedicated application(s). Those who provide mission-critical applications are obviously reluctant to have WAP gateway performance depending on other's applications and might prefer obtaining a private gateway.

Bluetooth Security Issues

Because Bluetooth only provides security for devices on the lower protocol layers, it would be possible for someone who steals your device to continue using it as if he or she were you. This lack of user-level security should be remedied by application-level security for individual applications. You can either use WTLS for WAP applications or TLS for those who have a TCP/IP stack. Depending on the sensitivity of the application, you can then complement this security with login procedures and/or other identification methods, such as iris scans and fingerprint readers.

Redundant Security

Just as often as we find applications that have a lack of security, we find those that have excessive security functionalities. Most of the time, this situation is not the fault of either the operator or the application developer; rather, it is a result of the nonexistent standard for generic security functions. You will find it common, for example, that the SIM card will authenticate you toward the GPRS system when you turn the phone on, and then the WAP gateway will do the same. Finally, you have to type a password in order to access an individual WAP application. Here, we see that the problem is pretty difficult to solve, because we want to keep the modularity (WAP can be used with other networks than GPRS) and diversity among applications (developers have to have the right to establish relationships with users). Maybe we can solve this issue with the advent of the full-blown service networks that come with UMTS, where the Common Directory (in PSEM) holds subscriber information that applications can access. The mobile IP architecture, with an AAA server as a regular IP node, might possibly help out as well.

Making Decisions and Security Perspectives

When making decisions for the applications, you of course need to look at the big picture. The following paragraphs offer some generic advice that you can use as guidelines when choosing security features for applications:

- Is the application security centric? In other words, does the bare existence of the application depend on how much security you can offer? In that case, you should perform the security evaluation first and then tailor the rest of the application around this evaluation. Building security into the product from the start and making all of the decisions on the way with this security in mind always helps you obtain the highest level of security.
- In other cases, you should gain an understanding of how the different parts of the protocol stack contribute to the overall security picture (for example, GPRS, Bluetooth, WAP, Java, and so on) and evaluate the result. Then, make decisions based on whether the security needs to be complemented with additional software. There are many security products available that can fill potential voids.

The operator has to make some additional choices regarding the security field. Some might choose to become CAs and provide certificates for those developers who are interested. Building a highly reliable brand and reputation would make developers less keen on getting their own gateways, although such highly secured gateways are likely to be available for use at a premium. As we stated previously in this chapter, the operator has a unique position in that he or she controls both the client (SIM) and the network (AUC) security in GSM-type networks. This situation has led to some interesting applications solutions being born in Europe's *second-generation* (2G) network. An example is the application where users can buy movie tickets over their mobile phone. The operator then charges the user for the ticket and communicates the transaction with the movie theater and deducts the cost from the user's phone bill. The movie theater then sends an electronic ticket to the cellular phone, which appears at the theater by using infrared or Bluetooth (see Figure 12.5).

The interesting part here is that the user does not have to leave a credit card number and is charged for things with the phone bill. All in all, security is about sufficiently good technology and trust. You can always make things more secure, and no solution is perfect. The application that uses client and server certificates that have to be signed with 40-digit passwords that change pseudo-randomly every day is probably pretty safe, but who would want to use it? Some security solutions, such as the credit card company-initiated SET, have had problems with gaining acceptance because people find them too complicated. The application developer should always have these factors in mind. With the new usage patterns that the Internet has created, people are always more reluctant to use a site that requires a username and a password. With the mobile Internet, where the input mechanisms are limited and people will get used to quick and straight-to-the-point applications, every barrier will reduce the number of users (provided that they still feel that the site is secure).

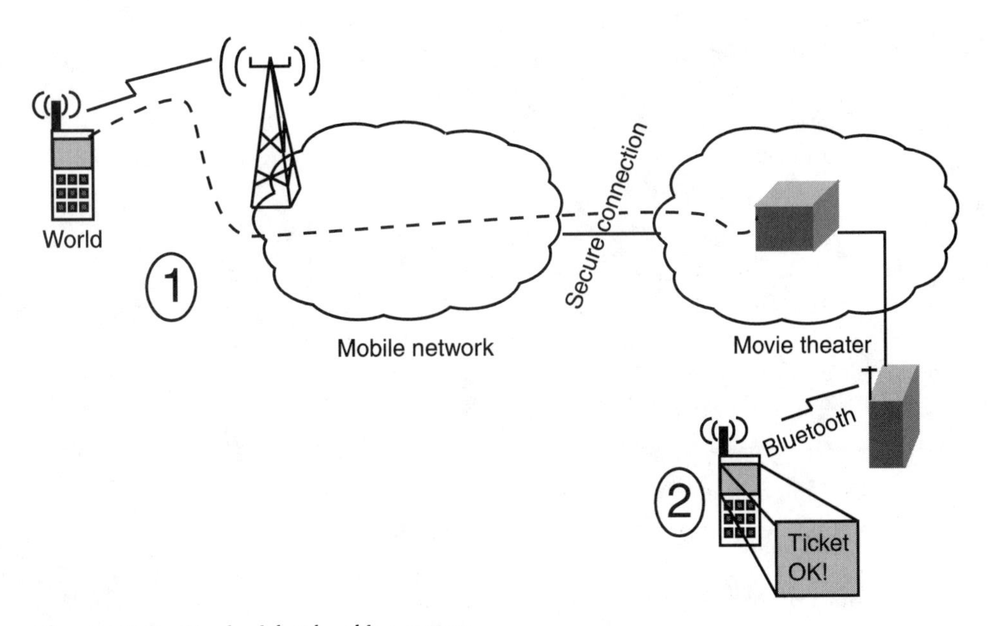

Figure 12.5 Movie ticket booking system.

Summary

The security mechanisms for mobile Internet applications are important to many developers. Because there are several layers in every application, each of those layers needs to be secure. A perfect lower-layer security can be worthless if the application spreads secure information generously. Security should, therefore, always be looked at from an end-to-end point of view, where every link in the chain from client to server is secure if needed. Because there are almost always usability tradeoffs (passwords, and so on) with secure solutions, the developer should always be very aware of how much security is really needed. Security involves perception and making the user feel safe (often the most important thing).

CHAPTER 13

Location-Based Services

Predicting the success of upcoming technologies has always proven very difficult, and not many have succeeded repeatedly. When you are trying to estimate the future acceptance of new technologies, it is common to perform market surveys that ask consumers how interested they are in taking advantage of these new technologies. This method often fails because users are more likely to like services that resemble things that they use today. Sometimes they are unwilling to believe in things that are unproven and that they have not seen before. We all know that consumers need time to get used to new things, and it takes time to gain wide acceptance of new technologies. If someone had asked consumers in the early 1990s whether they would like to buy groceries and books and perform their banking by using computer networks, not many would have been interested. In the same way, it is hard to anticipate what users will do with their mobile devices a few years from now. Every marketing survey, however, shows that there is one feature of the mobile Internet that users are keen to start using: location-based services. With location-based services, we take the personalization of services to a new level and tailor the applications to the physical position of the user. In order to fully leverage location-based services in wireless applications, however, we need to know about the technologies used and learn how to access them (through APIs, or Application Programming Interfaces).

Overview

The drivers for location-based services are a bit different in different parts of the world. In the United States, the government has set regulations that incoming emergency calls should be tracked by using the caller's mobile phone. This regulation is called *Emergency 911* (E-911) and specifies the rules with which mobile network and terminal manufacturers need to comply. The deadline for this regulation is set for October 2001 and forces the industry to speed up the implementation of positioning technology.

In Europe and Asia, the main driving force has been the urge for more advanced personalized services. Some operators started with location-based services even in the late 1990s, with a significant ramp-up in the first years of the twenty-first century.

Different needs have spurred different approaches toward achieving the goal of positioning users. In the United States, we primarily want dispatchers to extract our location information when we call 911. The requests would then be stochastic, and we would expect no significant load on the positioning equipment (it is not likely that one user will call 911 several times per day), which sets lower demands on scalability. On the other hand, the reliability is crucial, because such a system *has to work* when someone needs it. Although the U.S. market is likely to benefit greatly from the positioning features in the implementation of commercial services as well, the pace is set by the E-911 requirements and consequently it also decides many of the implementation issues.

The commercial forces in the rest of the world have placed pressure on the development of standards for positioning, and as wireless applications developers started to emerge, these commercial forces joined in the effort. Standards, however, tend to take time to develop, and some proprietary solutions have emerged in the meantime. Most of these solutions specify the actual positioning technology (such as how base stations calculate a user's position), but some go one step farther and unify the different positioning technologies in an API that the developer can use. In late 2000, Ericsson, Motorola, and Nokia joined forces and created the *Location Interoperability Forum* (LIF) in order to extend the focus on this standardization. The aim of this forum is to explore the complexity of positioning users who use different technologies (such as GSM, cdmaOne, TDMA, and their evolutions) and to facilitate the development of standards that ensure simplicity and interoperability. While the forum does not intend to develop its own standards, it will be active in solving the issues and contributing to the standardization bodies that are responsible for the different mobile systems (3GPP, 3GPP2, and so on). For more information, visit the forum's Web site at www.locationforum.org.

The industry seems to agree that a position solution adheres to the service network architecture, where applications can access a positioning API that a service capability server provides. In that way, the application developer would not need to know all of the details about the method that is used (or even about the mobile system). The *Mobile Positioning Protocol* (MPP) provides such an

API and is already used by most of those operators who offer location-based services. Location-based services that want to access positioning information from the *Mobile Positioning Center* (MPC) use MPP. The MPC works like a service enabler, and you can use the same positioning API regardless of the method that you use to position the user. We describe this solution in more detail at the end of this chapter after we look at the technologies that provide the actual location of a user.

Positioning Methods

In the following sections, we give some examples of positioning technologies—mostly to illustrate the basic thinking behind positioning. As we will see, many times the technology that we use only matters because of the accuracy that it provides. The applications developer can rarely affect the solution used, but this situation is in the hands of the mobile operator and the device manufacturers. The operator will have to consider how many changes the network can tolerate and how important it is to be able to position legacy handsets. More than likely, most operators will support both a network-based solution for backward compatibility and a handset-based solution that will increase accuracy. Through the provided service network API, the developer can access the positioning information that is available without having to worry too much.

The LIF specifies three levels of positioning (source: LIF presentation, fall 2000):

1. Basic level: location of all handsets including legacy devices (for example, CGI-TA)
2. Enhanced service level: location of all new handsets with improved accuracy and reasonable costs (for example, UL-TOA and E-OTD)
3. Extended service level: location of new handsets with high accuracy and higher cost (relative to category 2) with customer choice (for example, GPS)

When finding the location of a user, you must first ask the question, "Where in the mobile system should the positioning take place?" While a number of alternatives exist, they are generally either terminal based (the handset positions itself) or network based (the network locates the user).

Terminal-Based Positioning: GPS and A-GPS

The *Global Positioning System* (GPS) is a satellite-based positioning system that the U.S. government initiated and that is already widely used around the world. GPS is not only used for navigation and positioning (for instance, by ships), but it is also used for synchronization. cdmaOne and its migration systems use GPS receivers to synchronize base stations because this method provides high accuracy. Because people saw the GPS system as a significant advantage in warfare, the U.S. government added a small disturbance to the signals so that the highest accuracy level would only be available to the U.S. Army. This system of *Selective Availability* (SA) was then removed in May 2000 and enabled high accuracy for everyone. Consequently, GPS receivers now enable users to locate themselves with an accuracy of 5m to 40m, depending on the conditions (see Figure 13.1).

In order to position yourself by using GPS, you need a GPS receiver that receives transmissions from satellites. The GPS receiver listens to three or four satellites and can then calculate the resulting position or send the raw measurements to the network for processing. The satellites never need to consider how many users they are serving and where those users are; rather, they just broadcast the signal for the receivers to pick up. As the name indicates, the GPS receiver does not have to send anything to the satellites, which means less power consumption. Still, the computations that the GPS receiver needs in order to perform are significant if the processing takes place in the device (an important consideration for device manufacturers). Some manufacturers choose to integrate a GPS chip into the mobile device while others use a dedicated GPS receiver and a Bluetooth or cable connection in order to attach it. The use of GPS chips has started to pick up as size, power consumption, and price have gone down and paved the way for a more widespread use of this technology.

The most significant limitation of GPS has always been that that it requires a clear view of the sky. Consequently, it is likely that a mobile device that a user

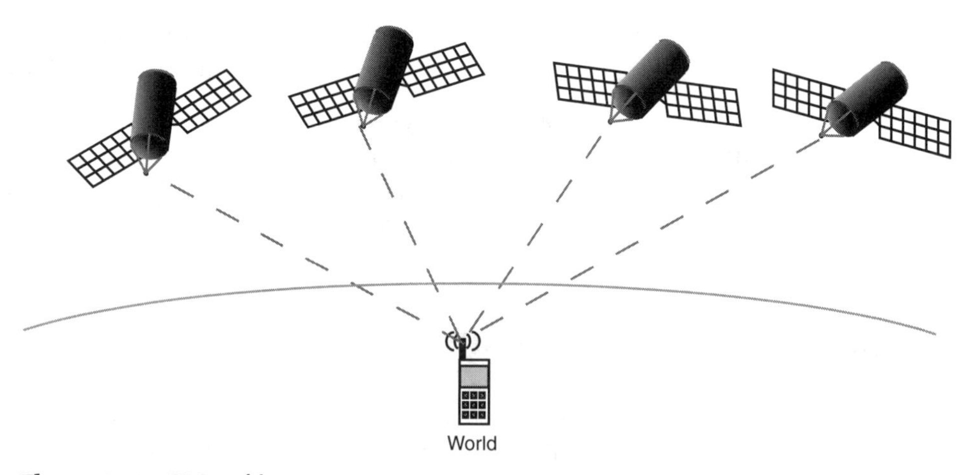

Figure 13.1 GPS architecture.

is using in a car will only be capable of using GPS if the antenna is located outside the car, which might reduce the device's usability. Car manufacturers are therefore looking into solutions where the mobile device is built into the car and attached to an external antenna. Another solution is to build the GPS antenna into the car and let whatever device the user has connect to it by using Bluetooth. Likely, GPS positioning solutions will also be complemented with a network-based solution, such as the cell identity, that does not need line-of-sight to the satellites.

When calculating position, the GPS receiver needs to know where the satellites are located, which imposes a delay on the delivery of the result. We sometimes call this delay *Time to First Fix* (TTFF), and this delay could be a burden for GPS as a positioning method. Common TTFF values are in the range of 20 to 45 seconds. If an application asks the MPC for a position and it takes up to 45 seconds to get the result, it severely limits the device's usability. This problem can be remedied in part by using the network to fetch complimentary data that the terminal can use for the calculation.

Network-Assisted GPS (A-GPS) uses network-based GPS receivers to help the terminal measure GPS data (see Figure 13.2). These receivers are placed around the mobile network in 200 to 400 km intervals and collect GPS satellite data on a regular basis. This can then be requested from the GPS-enabled terminal and enable the receiver to make timing measurements without having to decode the actual messages from the satellites. This process reduces the TTFF to one to eight seconds and makes GPS a much more compelling positioning solution.

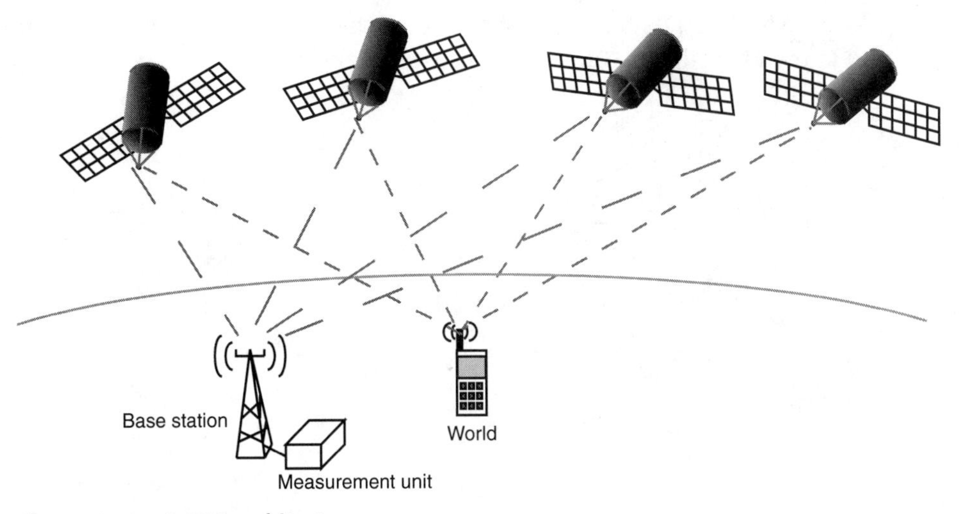

Figure 13.2 A-GPS architecture.

An interesting feature of GPS-based positioning solutions is that they enable user locating in three dimensions. For some specific applications, such as 911 rescue operations in the mountains, this feature might be of value because the rescuers can immediately see at what height the user is located.

Enhanced Observed Time Difference (E-OTD)

A concern with GPS is that it requires new hardware in the receiver, which is always something that device manufacturers are reluctant to include. Therefore, some argue that a software-based solution is preferable and is more cost-effective for the consumer. *Enhanced Observed Time Difference* (E-OTD) is a solution that calculates the time difference that it takes to receive data from different base stations and estimates the position based on that information (see Figure 13.3).

As we see in the figure, this system requires several (at least three) base stations to be in range of the mobile terminal and uses a triangulation method in order to calculate the result. For the measurements to be valid, however, the signals that are used for the calculation have to be sent at the same time (or the distances measured would have been measured at different times, and the user might have moved). In order for this process to work, an overlay network of *Location Measurement Units* (LMUs) needs to be deployed in order to provide an accurate timing source for the measurements. The E-OTD-enabled handset notes the time difference between the signals from the measured base stations. This time difference is then a measure of the distance between each of them, and we can use triangulation to calculate the position. Because the resulting position is measured relative to the base stations, the calculated position is relative to their positions. In other words, the base station coordinates need to be known in order to calculate the absolute position.

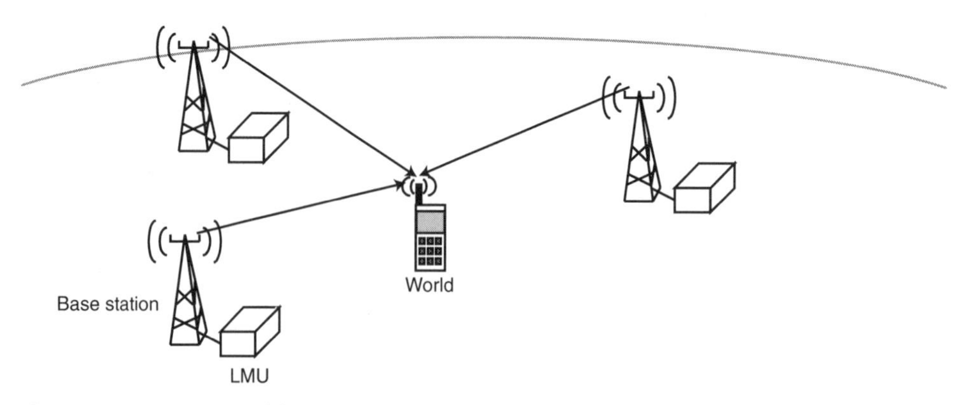

Figure 13.3 E-OTD architecture.

As with GPS solutions, the measurements are made in the terminal, but the calculations can take place in either the terminal or on the network. Again, making the calculations on the network makes the process less power consuming, but E-OTD still adds new requirements to the terminal. In order to use E-OTD algorithms both in idle mode and in dedicated/ready mode, the terminal needs to have additional memory, processing power, and battery power (compared to other handsets). The handset has to work harder during those periods of time when it otherwise would have been resting and saving battery power (idle mode). At this writing, it is unclear how many handsets will support E-OTD. GSM 03.71, Annex C, describes E-OTD.

Network-Based Positioning: UL-TOA

E-OTD uses triangulation based on downlink measurements in the mobile terminal, and which you can accomplish by using the network as well as the uplink. The *Uplink Time of Arrival* (UL-TOA) measures the received signal from a mobile station by using three different base stations. In addition to having an LMU that measures the time, UL-TOA relies on synchronized base stations. The synchronization of base stations is crucial and mostly takes place through GPS receivers or atom clocks in the base stations. Because cdmaOne/cdma2000 base stations are already synchronized from the start, UL-TOA is more compelling to use in these systems than the asynchronous GSM/TDMA/EDGE/WCDMA systems. Figure 13.4 shows that the three base stations all measure the time of arrival of signals from a mobile station and then calculate the signals' positions in a manner similar to E-OTD (but this time, on the network).

Calculating the absolute position again requires you to know the position of the measuring nodes. Because this method is only located on the network, it works

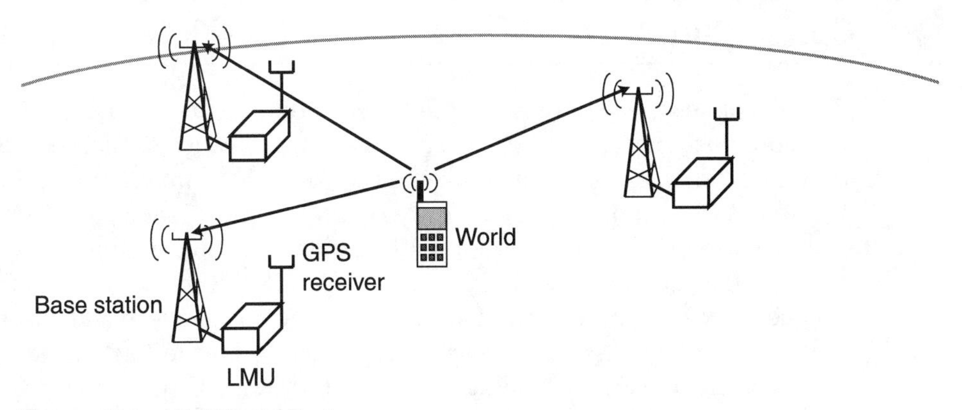

Figure 13.4 UL-TOA architecture.

with all handsets—even with those that never were designed with position features in mind.

You can expect the accuracy of UL-TOA to be between 50m in urban areas where base stations are densely deployed and 150m or more in rural areas.

CGI-TA

UL-TOA requires additional equipment to be installed on the operator's network, which some operators might be reluctant to do. An even simpler method of positioning by using the network is to look at the cell in which the user currently exists. This information is already available on the network; therefore, you do not need any network add-ons. We commonly call this method *Cell Global Identity* (CGI). The granularity of CGI, of course, depends on the cell size but usually is sufficient for proximity services (where is the closest restaurant, for example). Depending on which configuration you use, the resulting positioning area (the area where the user is located) is either a circle (an omnisector antenna configuration) or an approximate circle sector (when three sectors and directional antennas are used, as seen in Figure 13.5).

In order to take even more information into account that is already available on the network, you can use the *Timing Advance* (TA). TA is a measure of how far away from the base station the mobile user is (and therefore, you can shrink the uncertainty area of the positioning). The same configurations as in Figure 13.5 with CGI-TA appear in Figure 13.6.

The accuracy for CGI-TA is in the range of 100 to 200m, which is a very good result for such a simple method that works with legacy handsets.

Which Solutions Will We Use, and What Are the Consequences?

We will implement all of these solutions many times—in some instances, applying more than one solution to the same system—because of their complementary features. An excellent example is a network where we use A-GPS for those handsets that support it (when we are outdoors; line of sight to the satellites) and then use CGI-TA as an indoor method to fall back on and for users who do not have GPS receivers. A-GPS seems to stand out from the rest by providing excellent accuracy even in rural areas, but on the down side, it adds the most to the cost of the handset. Those systems that can position low-end and legacy handsets are always appealing, because they enable instant mass-market access.

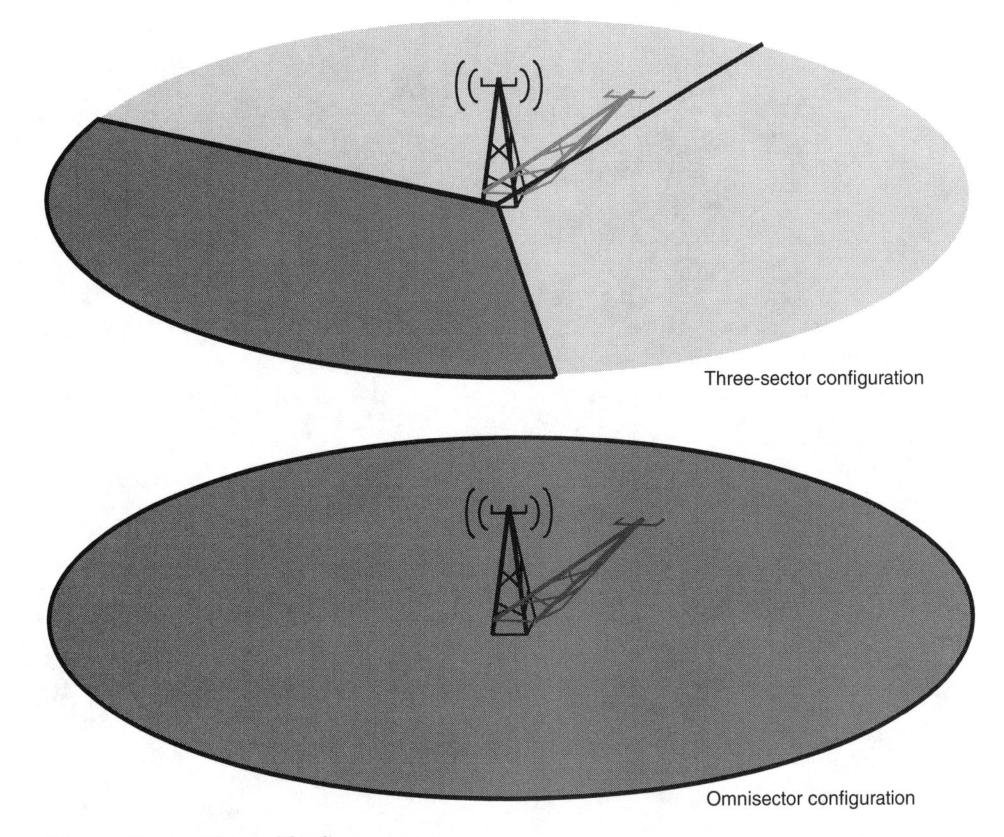

Figure 13.5 CGI positioning areas.

Because the positioning functionality is abstracted from the developers and (most of the time) from the user, the available technology does not make much of a difference to developers. Of course, some applications will rely on the highest level of accuracy for fully featured operations, but even then a solution that we can fall back on must be acceptable. The main aspect apart from that is the time that it takes to achieve the position. This time consists of two parts: the delay to get to the positioning center, and the time that it takes for the center to determine the desired position.

Because the positioning server is likely to be part of the service network, the developer needs to ensure that the application server that hosts a location-based application is located in such a way that it can communicate with the positioning center as quickly as possible. If you use HTTP over TCP/IP in order to achieve the position, the famous handshakes will make enough delays themselves so that the link (at least) will be fast. If an application server has a ping rate of 500ms just to get to the positioning center, this delay will be noticeable to the user.

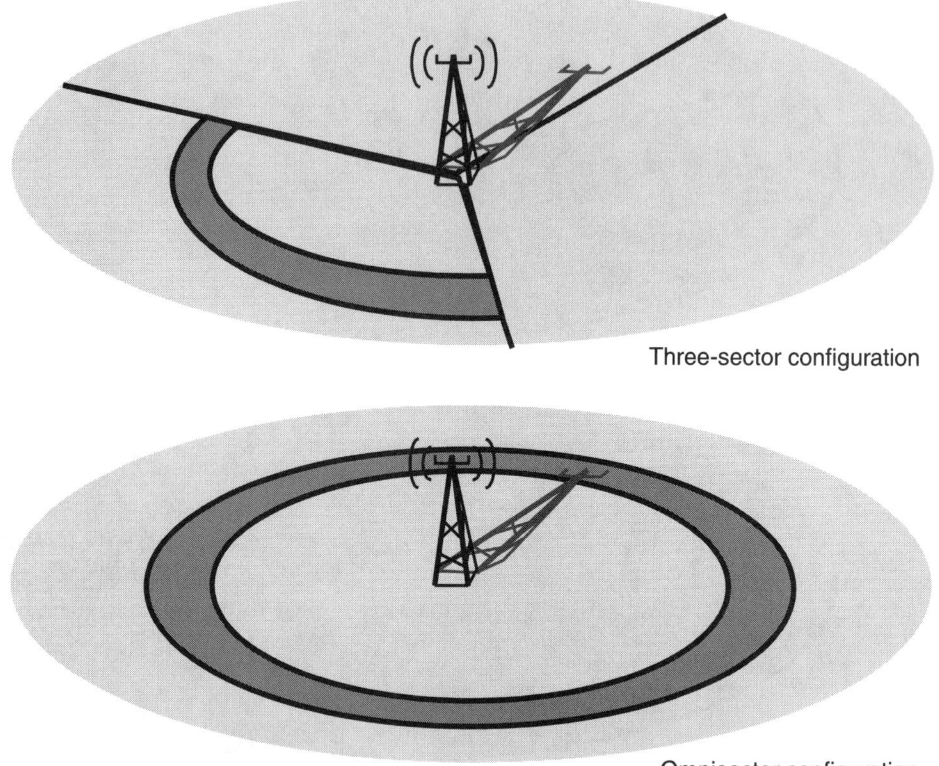

Figure 13.6 CGI-TA positioning areas.

In addition to the communications delay, some positioning methods take time to deliver the result. For E-OTD, for example, location processing times (the delay before achieving the requested position) have been measured at three to five seconds, and those figures are reported by the most convinced evangelists of the technology. A GPS system without network assistance can need as much as 45 seconds of processing time, while A-GPS shortens that time to one to eight seconds. For UL-TOA, the delays should be in the same range as E-OTD (a couple of seconds). Keeping this factor in mind when designing applications leads us to minimize the calls to the positioning center (and possibly designing it concurrently with other tasks).

Security and integrity protection will be crucial for the success of location-based services. People want the extended services but are reluctant to let everyone know their location. Everyone in the industry seems to agree on this statement, and users will be able to control when their position is available and how. As with most security aspects, it is all about making the user feel confident in the technology and trusting the operator and application provider.

We can clearly illustrate this point when the integrity issue often leads to heated debates, although the cell identity of a user is already available on existing networks.

As we mentioned in Chapter 9, "Application Architectures," the positioning functionality will likely be located on the service network for most operators. Most of the time, the application server-side of an application will make function calls to the positioning center, but the interface is usually quite generic. You can access it conveniently from other nodes, as well.

Example of Positioning API Usage

There are two main methods for accessing positioning information, and the LIF is the driving force behind standardization on both tracks:

- Terminal-based API, where an application that is running on the device accesses the positioning technology hardware/software
- Service network-based API, where an application server or anyone that has the rights can fetch a user's positioning data

All of the early entrants to the market have used the second method, because they can perform this task independently of the handset and of the positioning technology being used. It should be noted, however, that terminal-based positioning always needs to be supported in the terminal, as it will be required to report the measured position data back. The following section explains how the MPS *software development kit* (SDK), which is included on the accompanying CD-ROM, illustrates the thinking behind developing applications that use a server-side positioning node.

The development kit contains the following items:

- A Java class library
- Java example programs
- A protocol emulator, which is a local test server that implements interface stubs of the real MPC and can return realistic positioning information
- Visual Net, a tool that you can use to build mobile networks for testing purposes
- User guides
- MPP 3.0 documentation

The application accesses the MPC by using standard HTTP, as illustrated in Figure 13.7, and Web browsers and Java clients are examples of access methods. In the figure, a mobile client is shown (but it could as well have been a desktop PC or an application server).

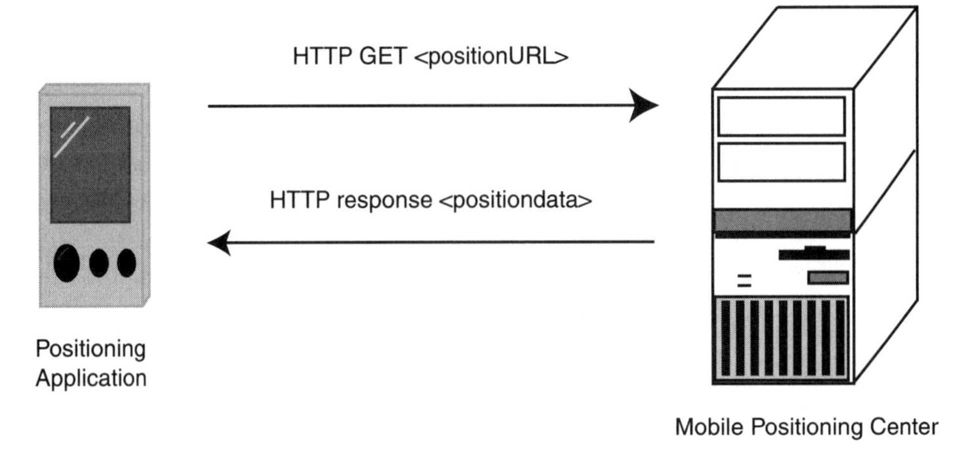

Figure 13.7 Positioning request/response.

In the request, nothing is mentioned about the underlying mobile network or positioning technology that is used. The requests can be complemented by TLS (SSL) security in order to ensure privacy.

The format of a request is shown as follows (an ordinary HTTP **GET** command to the MPC):

```
http://position.on.demand/PositionRequest/Direct?<ParameterList>
```

The parameter list is a set of parameters and their assigned values, separated by an ampersand (&):

```
Parameter=ParValue&Parameter2=ParValue2 . . .
```

An example request could be as follows:

```
http://my.favoriteMPC.com/PositionRequest/Direct?USERNAME=JohnDoe&PASSWO
RD=whatever&POSITIONING_ITEM=55512345678&POSITIONING_TIME=(time+now)
```

This request asks the MPC with the address my.favoriteMPC.com to return the positioning of user JohnDoe with phone number 55512345678. The POSITIONING_TIME item specifies that the positioning request should be effective immediately. The SDK on the CD-ROM includes a complete list of available parameters.

The response from the MPC can then look something like the following:

```
<Head RequestID=2.873191662.1 AnswerID=1>
<MS=46555303132
        RequestedTime=19971015134510+0200
        Error=0
        GeodeticDatum=WGS-84
        HeightDatum=NotAvailable
```

```
          CoordinateSystem=LL
          PositionFormat=IDMS0
          <PositionData
        <PositionArea
               Time=19971015134510+0200
               <Area=Arc
                      <Area=Point
                             Latitude=N600920
                             Longitude=E110808
                      >
                      InnerRadius=20009
                      OuterRadius=21109
                      StartAngle=300
                      StopAngle=60
               >
        >
        <PositionArea
               Time=19971015134510+0200
               LevelOfConfidence=100
               <Area=CircleSector
                      <Area=Point
                             Latitude=N600920
                             Longitude=E110808
                      >
                      StartAngle=300
                      StopAngle=60
                      Radius=21109
               >
        >
      >
   >
   <Tail RequestID=2.873191662.1>
```

In this example, a GSM system with CGI-TA is used for positioning. The rows of the response should be interpreted in the following way:

MS = 55512345678 (indicates the phone number of the mobile that has been positioned)

RequestedTime = 20001120102724 (says that the positioning time requested by the user was 24 seconds past 10.27 on November 20, 2000; the +0200 indicates the GMT difference in hours)

Error=0 (indicates that the request did not return any errors)

GeoDecticDatum=WSG-84 (used to describe the format of the position used)

Heightdatum=NotAvailable (states the lack of vertical positioning information; would require GPS)

CoordinateSystem=LL (defines the coordinate system used)

Positioningformat=IDMS0 (specifies the output format of the geographical position)

PositioningData (signals that the actual positioning information will follow; because this mobile has not had a handover since call setup, two positions can be delivered; described by two entries of PositionArea)

Time=20001120102724 (specifies that the time when the positioning was actually performed was the same time that the user requested)

Area=arc (with its parameters, draws the area within which the user has been located by using the TA value; the area is used to describe the uncertainty of the measurement)

The second PositioningArea (only based on the serving cell, and sometimes this area is the only area that can be delivered)

LevelOfConfidence=100 (states that the user is within the specified area with 100 percent probability)

You can use the included Java libraries when you are integrating the positioning requests into any application, and most experienced programmers will feel comfortable and familiar with this environment. The HTTP request format also makes it possible for servlets to access the information, provided that they know the username and password. This situation often makes it crucial for the developer to work closely with the operator in the later stages of implementing a positioning application.

Summary

Positioning definitely will be one of the most important features of the mobile Internet. The technologies can be either handset based, such as GPS and E-OTD, or network-based, such as UL-TOA and CGI-TA. GPS is the most accurate method but is also fairly expensive. CGI is the cheapest and easiest method to implement but has low accuracy in rural areas. The developer should not have to worry too much about how the different technologies work; rather, he or she should develop for an open API that hides the details and that works independently of the underlying positioning technology.

CHAPTER 14

Testing the Wireless Applications

When you are starting a new and exciting project that has great prospects, your first consideration is probably not a test strategy for the application. All too many developers start thinking about test strategies when they have completed just about everything else, and everyone is eager to get products on the market. Those who start with some tests early in the development process can benefit from valuable lessons learned and benefit from this knowledge in all of the following steps of the process. The most difficult part of testing wireless applications is probably realizing what kind of testing you need and how to perform the testing. You must complement the traditional test tools that Web designers and software programmers (C, Java, and so on) use with tools that test the wireless network properties and the target devices. Once the developer knows about the needs and the available tools, testing is no longer a hassle but is instead a valuable help in ensuring quality, interoperability, and even a faster time to market.

Why and How to Test

The mobile Internet is developing at a raging speed, and just about everyone is struggling to keep up with the new technology. New devices emerge on a weekly basis, and more and more players are entering the market. When the WAP arrived on *second-generation* (2G) networks, the first developers only needed to make sure that the applications worked on a handful of devices. There were just a few devices for each of the mobile *operating system* (OS) platforms (EPOC, Palm, and WinCE). The number of available, supported devices will definitely not become smaller as we move farther into the mobile Internet future. On the contrary, as more and more complementary technologies such as the *Global Positioning System* (GPS) and Bluetooth become pervasive, there will be a much larger array of devices to support.

This situation probably means that applications will not only have to work on a variety of target platforms, but that a competitive advantage will also arise for those devices that have different presentations on different form factors. A WAP application that just shows basic text content on a mobile phone with four rows can show additional pictures and rich content on a larger *personal digital assistant* (PDA). In addition, an application in today's world should, of course, work nicely with devices that come out months and perhaps even years later. The key to this functionality is, of course, following standards as much as possible but also testing the application on as many available devices and device emulators as possible. Manufacturers of new devices not only take these standards into account, but also consider the de facto standards created by existing devices in order to ensure that the available applications will work. In other words, a developer can have a much bigger chance of ensuring compatibility with an upcoming device if he or she tests the applications on emulators and on real devices that are currently on the market. Consequently, testing on device emulators and real devices is crucial.

In addition to testing how the application will perform on target devices, we must make sure that it will run on the networks for which it is aimed and on networks that we will probably introduce in upcoming years. This method is the only way that we can ensure that the varying and sometimes harsh mobile network conditions will not affect performance.

In addition to the regular function testing that we use for any software product (function testing and so on), we typically add the following tests for a wireless application:

Graphical User Interface (GUI) and usability testing. This testing involves making sure that the application looks nice on a wide number of target devices and that it offers a user-friendly interface. You should perform this test by using emulators from the start of the development process, and perform this test on real devices as the product becomes more mature.

Network performance testing. This test involves making sure that the application performs well even in the harshest of conditions, such as passing under a tunnel or coming back from periods without coverage. Getting some of this feedback in an early phase can drastically improve the end performance of the final product.

Server-side testing. As we implement more advanced functionality on the server side, we must test both the functionality and robustness of that end.

This testing includes the application server that hosts the server side of the application and also other service network components that are involved, such as positioning servers and the WAP gateway.

When we go through these tests, the first question is how to find the right test tools. This question raises the issue of whether to use emulators or to test directly on real devices and real networks.

Emulators and Real Networks and Devices

Not only do we determine the testing environment by the actual test tool that we will use, but also by the target devices and networks. Testing everything at once in a full-blown, immediate manner is not only impractical and expensive, but also very difficult. On the contrary, the testing should start by using emulators and *Graphical User Interface* (GUI) testing, preferably performed separately from network emulator testing. In that way, you can focus the testing on one thing at a time and isolate potential issues. Because GUI emulators (phone/PDA emulators) are often freely available on the Internet, the most common way is to first concentrate on testing the functionality and user interfaces offline before starting to optimize the wireless properties. You should perform the testing in several phases, however, so that you follow a first round of GUI tests by simple network emulator tests and then repeat the same process until the result is satisfactory. This process enables you to discover big flaws early on, and many times a testing session can be a very valuable workshop for developers (see Figure 14.1). This statement is true not only for Java and C applications, but also for WAP and other thin client services.

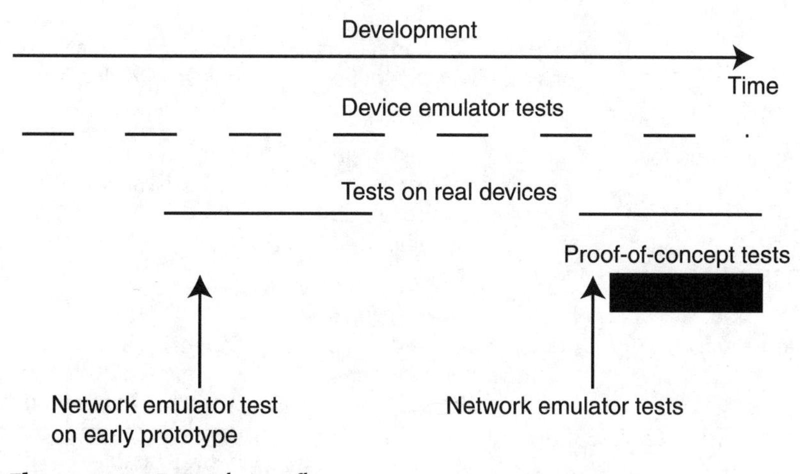

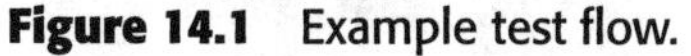
Figure 14.1 Example test flow.

Apart from enabling the test of each part separately, this way of testing also leverages the advantages of using both emulators and real networks and devices, as we can see in Table 14.1.

Real versus Emulated GUIs (Devices)

Apart from the obvious aspects, such as an emulator being cheap but not always 100 percent correct, there are a number of important advantages of using an emulated device. Some very interesting complementary test tools are starting to pop up that enhance testing. One example is WinRunner by Mercury Interactive, which uses a test GUI with the device emulator in order to test multiple executions of a task. This functionality enables the developer to record a set of clicks on the PC-based emulator and then let WinRunner execute this test scenario hundreds of times in order to measure performance and stability.

Real versus Emulated Mobile Networks

With the fast pace of network development in the mobile Internet world, developers often create an application for a network that is not widely available in

Table 14.1 Different Test Methods

TEST METHOD	ADVANTAGE	DISADVANTAGE
GUI (device) emulator	Free (mostly) You can quickly test many devices; can be used together with other test tools for repeated testing; tests can easily be logged	Usually the speed is not the same as the real device; other functionality might also differ slightly.
Real device	The exact look and feel as the consumer will experience	Expensive; not always available to developers before launch
Emulated mobile network	Inexpensive to test on; easy to repeat test scenarios; logging of results available; easy to invoke advanced features, such as interruptions	Properties may differ from real networks
Real mobile network	The exact look and feel as the consumer will experience	Expensive and hard to test on; might not be available when the application is being developed; exact scenarios not repeatable

the area where the developer is located. Even when the network is available, it is difficult to use the network for more than a final proof-of-concept test. With a real network, you cannot repeat the exact test scenario two times because of the many factors that affect the performance of a real network. Even if you had your own network in a laboratory, the radio waves travel very different paths if your position is just slightly different. Add to that factor the number of other users on a typical commercial network, and you will find that it is crucial to first optimize the application by testing it on an emulator and then verifying the performance on real networks.

With an emulator, it is easy to record a test scenario and then repeat that scenario later in order to determine whether the developer has managed to improve the performance. You can easily add advanced features such as interruptions, different operator settings, and high-end handsets to the tests in order to make sure that the application is robust and ready for a variety of networks (and not just for one or two networks that have associated handsets).

To summarize, you should perform the testing mostly by using emulators, but you should always verify the testing on real networks and real devices. Now, let's look more in detail at how we can perform these tests.

GUI and Usability Testing

While the majority of developers perform their GUI and usability testing in their own labs, some public test sites are emerging. This section describes some of the basic factors to consider when performing these tests (regardless of who creates them).

Although many quickly find a favorite emulator or a favorite device, you must be open to the needs of consumers. The more testing that you perform on a variety of emulators, the lower the risk of losing entire market segments that comes from the application being incompatible with some devices. The time consumption of such tests is, however, a significant obstacle in the stressed development process. We can remedy this concern by choosing a main development platform with an associated emulator and then performing tests over most available emulators at certain milestones in the development process. The style guides that usually accompany the emulators are a great help with getting the most from the platform in question. On the CD-ROM and Web site that accompanies this book, there are a number of good emulators available that cover a range of devices and that you can use as a starting point. All of the key players in the device and operating system business are keen on having cutting-edge developer tools and should always be important bookmarks for the dedicated developer.

While you are making the application actually work on the desired devices, you should know that this phase of the testing should also include basic usability tests. This testing is especially important for browser-based applications such as WAP. While users commonly use the Web browser for surfing, the mobile Internet involves straight-to-the-point information. How much information can the user access within 10 seconds on your site? When you think about such things, you must picture the user as someone who is totally unfamiliar with the application and not the experienced tester or developer within your company. A useful exercise is to draw a site map that shows the tree of the decks in the application, as shown in Figure 14.2. In the site map, we show the percent of the test users who access a certain deck.

Figure 14.2 illustrates the WAP site of a small movie theater. The main deck gives the user three choices: book tickets, look at movie information, and check out information about the theater. In this example, we can see how the current movies are fairly far down in the hierarchy, but lots of users still access them. As a result, the average user has to perform many navigational steps in order to reach the desired information. Here, it would be easy to move this current.wml deck to a higher level by making it a choice in the main deck, as shown in Figure 14.3. While this example is pretty small, it makes a lot of difference if we can make this kind of change to larger sites.

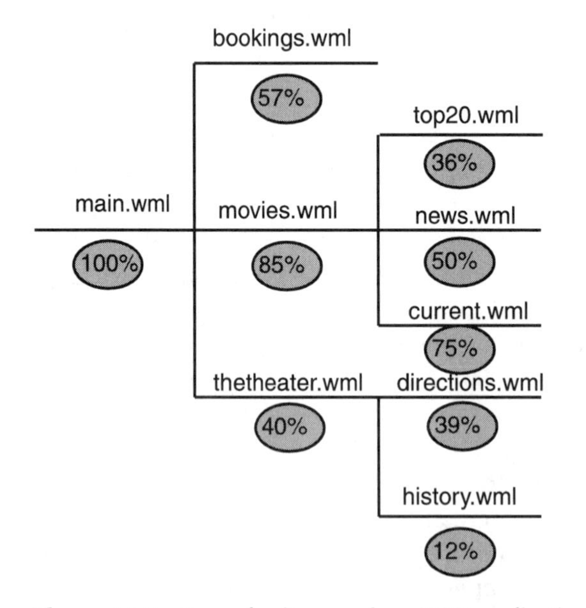

Figure 14.2 Example site map for a WAP application.

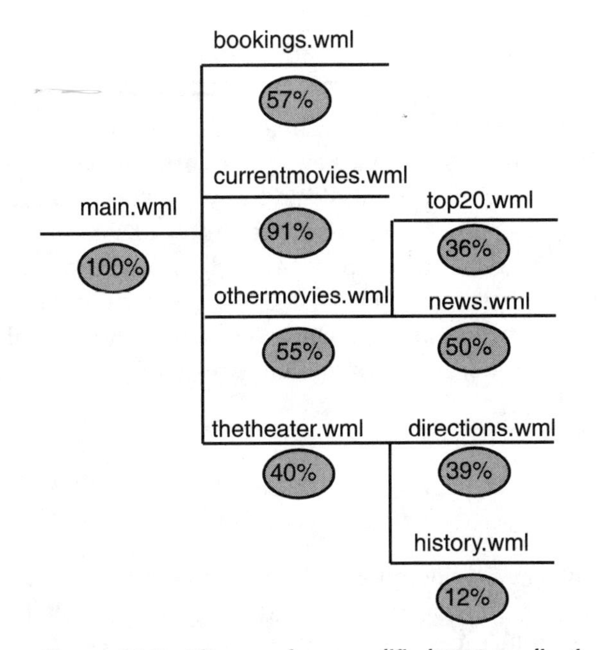

Figure 14.3 Site map for a modified WAP application.

In the modified site, we have moved the deck that shows the current movies to a higher level; consequently, it catches a lot of hits. This kind of testing improves the usability as well as the overall perceived performance.

We should investigate some generic GUI aspects, such as the following:

- How much navigation does the user typically have to go through? Mobile users are more inpatient than their fixed-Internet counterparts.
- Is user input minimized? Typing on a wireless device is not as easy as on a full-size keyboard.
- Can the screen of all target devices handle the content? The refresh rate of devices can vary slightly, and animated content might be affected.
- Does all of the GUI code work on all target platforms? If not, how much do we have to change, and how can we avoid this situation?
- Is there support for international characters, or are you 100 percent sure that no one in Germany, China, Sweden, or Finland will use this application?

You should repeat the GUI and usability often and at certain milestones on a variety of target devices. Because the performance of emulators and real devices often differs significantly, you must also run tests on the real platform early in the process.

Network Emulator Testing

While the new GUI aspects might present a significant learning curve for developers, you will find it more challenging to take the network aspects into concern. Even those who have studied this book and have absorbed all of the advice (and maybe even developed for wireless before) will need to test it in order to make sure that the application works properly. The key is to get initial feedback as early as possible in the development process in order for the development team to avoid future mistakes. Once the development is almost complete, you can perform another round of emulator tests followed by proof-of-concept tests on real networks.

Wireless Emulators

Having been in teams that develop detailed network simulators for research and capacity estimate purposes, I have seen how utterly complex such a product can be. Development usually takes place so that we examine one feature at a time (for example, handover, power control, and so on). Looking at the resulting complexity, we can see that apparently, an emulation of a complete *General Packet Radio Services* (GPRS) or *third-generation* (3G) network would be almost as complex as the real thing. To avoid this situation, we would have to carefully examine the purpose of the emulator. We can then develop an emulator solely for the purpose of testing mobile Internet applications and then design it accordingly. The result is a product that emulates GPRS and 3G from the application's point of view and that concentrates on those aspects that are relevant to the applications.

The *Mobile Applications Initiative* (MAI), www.mobileapplicationsinitiative.com, uses an emulator that reflects this thinking and exists specifically for application testing. The *Global Applications Test Environment* (GATE) was initially developed as a GPRS test environment but now also includes 3G network technologies. GATE is connected to the application where the live network would have been and emulates it, as we can see in Figure 14.4.

In other words, any application that runs on top of the *Internet Protocol* (IP) can connect to the GATE—and we can then investigate its transmission properties. The connectivity commonly takes place via regular Ethernet networks, but you can also use serial cables, wireless *Local Area Networks* (LANs), or Bluetooth when attaching the terminal. In the early phases of application development, an emulator that runs on a desktop PC (connected to GATE via Ethernet) is a convenient way of starting the testing even before you have completed the full device testing.

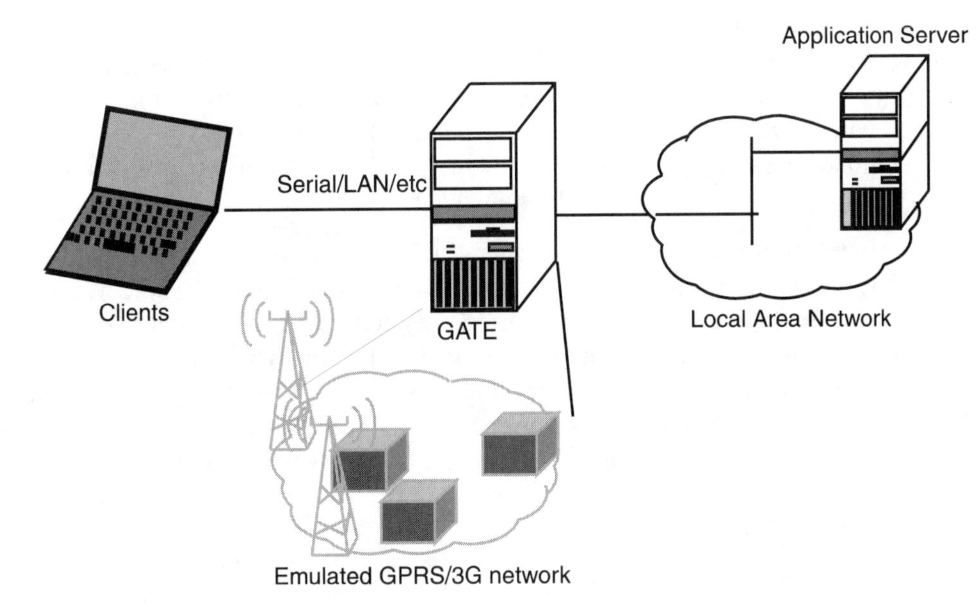

Figure 14.4 GATE setup.

We can then view the properties in the GATE GUI and view and change characteristics such as latency, interruptions, background users, and terminal capacity. Figure 14.5 shows GATE 2.3 when it is emulating a typical GPRS network.

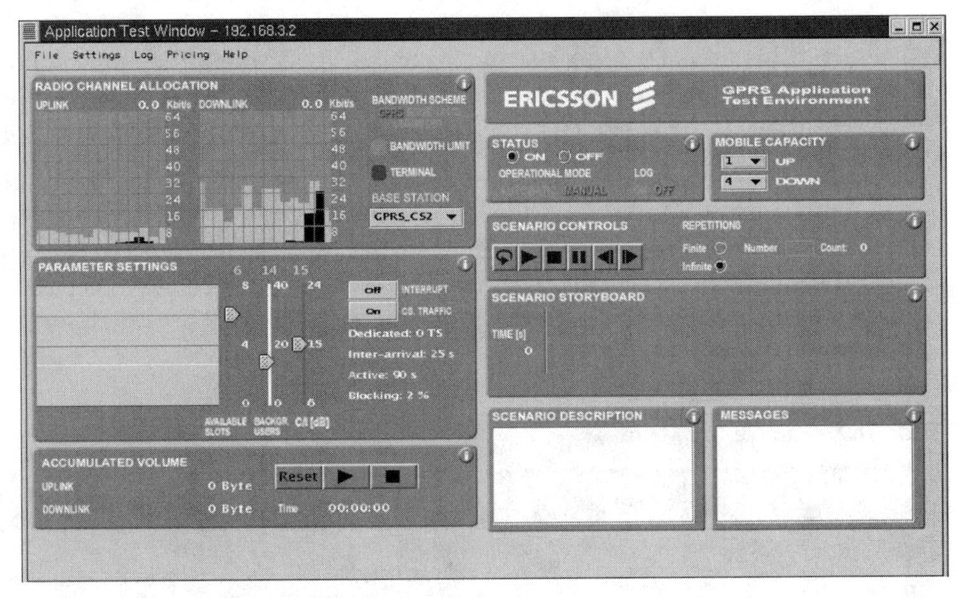

Figure 14.5 A GATE GUI with parameters.

By using this interface, the tester can put the application in radio shadow (interrupting the transmission) or double the number of users in the cell with a mouse click. There are also logging features and a throughput recorder that you can use to measure the use of bandwidth over a longer period. This feature makes it possible to find potential problems quickly in the application and optimize products for the wireless environment.

In the future, there might be other emulators that have similar capabilities, and it will be up to the developer to find an appropriate tool. The GATE is not for sale, and tests only take place in the 25-plus MAI labs around the world. While this situation might initially seem like an obstacle, we can only achieve the full benefit of wireless network testing when we perform the tests with experts—the application expert (you, hopefully), and the network expert (here, we refer to the MAI personnel). When application and wireless expertise unify in this way, the testing session becomes more of a workshop with mutual knowledge transfer (rather than a terrifying judgment day).

Test Cases and Wireless Scenarios

When testing the wireless properties of an application, we must know both aspects well: how the application should behave, and how the networks are likely to affect the application. Commonly, the network expert does not have a clue about the application behavior (and vice versa). Therefore, we should combine the test cases of the application with the wireless scenarios that we expect. An example of this thinking appears in Table 14.2, where rows represent test cases and the columns illustrate the wireless scenarios. We can then limit this test matrix to just those entries upon which the application expert and the network expert agree. Note that in this example, we only use three test cases and three wireless scenarios, while a real test typically consists of more.

As a result, both parties contribute what they know well, and the results mirror the joint view of the application. The criterion for each entry varies depending on the severity of the conditions. While the application should respond within a specified time in a medium-bandwidth test scenario, just surviving a 20-second interruption might be enough. For some test scenarios, we do not even expect

Table 14.2 Sample Test Matrix

	INTERRUPTION (5 SEC.)	LOW BANDWIDTH	IRREGULAR BANDWIDTH
Fetch Inbox			
Send mail A			
Send mail B			

the test case to finish; instead, the perceived performance (user control and user information) should be good enough.

Some examples of wireless test scenarios for GPRS specifically include the following:

Low bandwidth. Can the application still perform tasks when data barely gets through? This scenario is interesting when you are measuring overhead, because protocols such as the *Transmission Control Protocol* (TCP) have to struggle here with retransmissions that timeouts trigger. GPRS networks especially must be capable of handling low bandwidth, because the cells are very crowded during peak hours.

Interruptions. These are probably the most important test scenarios and the ones that most developers fail. Interruptions are not only performed with varying duration, but also during different parts of a transaction. Sometimes an application can survive even the longest interruption as long as it manages to get a request through to the server (while a tiny one at the time of the request will make the application very sad).

Varying bandwidth. The only thing about the bandwidth of GPRS that a developer can take for granted is the uncertainty. Voice users are likely to be prioritized by most operators, and getting a few new voice or circuit-switched data users into a cell will severely downgrade the throughput. Therefore, applications need to be capable of coping with this and need to adjust to bandwidth variations regardless of the direction.

Once you have found the application robust and fit for the wireless environment, you can perform testing with real devices on real networks. For those applications that interact with the service network (call control, positioning, and so on), you can either emulate this added functionality with *Application Programming Interface* (API) stubs or use the real servers. Please note, however, that the server part also needs testing attention.

Server-Side Testing

While the server side is naturally tested with various test cases during the function testing, there are aspects of it that require special attention (including application server robustness and response times, service network intercommunication, and overall load testing).

Because many of the server-side aspects are beyond the control of the developer, the task is more about finding out what the prerequisites are and making the best out of them. What kind of application server will we use, and what functionality will it provide? How much latency can we typically expect between the application server and the positioning center?

Application Server Issues

As we previously mentioned, the application server technology is rapidly evolving. The demands for multimillion user capacity and upgrades on the fly have led to much more developer-friendly platforms. Adding new functionality also means that we need to test this functionality, and this procedure is sometimes very difficult. The least that a developer can do is to start a dialogue with the provider of the application server (for those who choose not to buy their own) and get some answers to the most common questions:

- How is the server affected during hardware upgrades? While most application servers enable hardware upgrades to be performed while still offering service, we must find out how the performance will be affected during the process.
- If your application is running on a server and needs to be upgraded, how will users experience this situation? Can the service be uninterrupted, or are you forced to upgrade only during the night?
- If other applications crash/malfunction on the application server, are there disturbances in the environment of your application?
- Can the client and server interoperate even if the server side has been disconnected for some time? While these problems mostly are found during interruption tests of the network, there are some server-specific issues that you must consider. Sometimes the server keeps internal states for the users who might be affected if the server-side application or the entire application server needs to be restarted.
- What are the processing times on the server side for all possible requests and operations? Remember that you expect to have up to millions of users for your application.

Even those who are not planning to serve the application on their own servers should make sure that the platform in question is available for continuous testing. For those situations where each application needs additional databases and content servers, you must make sure that the architecture scales well and that you have tested its functionality.

Service Network Intercommunication

Mobile Internet applications become more and more advanced, and the evolving service network has contributed a lot to this situation. As always, adding more functionality also means that more testing is needed in order to ensure that it works correctly and that it can work together with all other parts of the application. Therefore, it is vital to understand how the service network will affect the application (and, if possible, how to perform tests that verify this functioning).

When there are new products that conform to a set standard, there are usually some initial slight differences between different implementations. This experience has been painful for WAP developers who have become used to different behaviors of gateways from different vendors. This situation is likely to be the case for other Service Capability Servers (SCSs) that are added according to the latest 3GPP standards. Therefore, it is important to understand those differences and also to verify that the applications can handle them all.

This is very difficult to test because the service network nodes and its underlying architecture are very expensive, and not many operators will let developers jeopardize existing services just to make sure that the new ones will work. Also, operators are more likely to adopt applications that have already been properly tested. As full-blown Service Networks become available with the advent of the 3G networks, we will see whether new ways to test these things appear. A simple method of testing is to use the SDKs that the manufacturer of the SCS in question provides. In those SDKs, an API emulator is usually available that developers can use.

End-to-End (Proof-of-Concept) Testing

Once you have thoroughly tested the applications on emulated and real devices, on radio networks, and on service networks, you must verify that everything works together in a live environment. Again, this test environment is extremely expensive, and your best bet is to find some dedicated test site that provides this service. In some cases, infrastructure vendors such as Ericsson and Lucent provide the service as part of their goal to provide end-to-end solutions to their customers. Developers will find it very convenient if someone else can take away this burden.

Although some of the tests appear difficult and cumbersome, it still is one of the most important steps in the development process. As applications become more and more advanced, operators will favor those vendors who constantly deliver applications that have been thoroughly tested. This situation might make testing migrate from being a competitive advantage to being a prerequisite for deployment. Because our mobile phones traditionally have been more reliable and robust than corresponding desktop PCs, the applications developers cannot design applications that compromise this stability.

Getting Help

Luckily, developers can get help during the development process, both in terms of tools and services. The online communities are chat rooms and mailing lists, where developers share problems and solutions. Those communities often also have corresponding *Frequently Asked Questions* (FAQs) that describe the most common issues and show how to get around them. The following list provides examples of some online resources that developers are using as of this writing (an up-to-date list of links are found on this book's Web site):

- www.anywhereyougo.com
- www.ericsson.com/developerszone
- forum.nokia.com
- www.motorola.com/developer
- www.wirelessdevnet.com
- www.mobileapplicationsinitiative.com

When you are testing the *Graphical User Interface* (GUI) of the application, you can always turn to the device manufacturers as an obvious resource, but there are others that even let you test WAP applications online. Examples of GUI testing resources include the following:

- www.gelon.net
- www.anywhereyougo.com

At this writing, the only site that offers help with network testing is the Mobile Applications Initiative at www.mobileapplicationsinitiative.com. Here, developers can sign up online for testing on the GATE tool that we mentioned previously, and there is probably a center in your neighborhood. Some operators and service providers also offer developer support and can often provide guides as to how applications are typically offered and implemented.

Summary

Testing has always been a part of software and web content development. The many complex technologies involved in the mobile Internet make it crucial to expand these tests to include the specifics of wireless networks and terminals. The three basic areas to test are the handset/GUI, network properties, and server access. In addition, all of these areas can be tested in both emulators and real environments. Both have their strengths and weaknesses and complement each other. This often leads to the need to partner with those who provide those tests and tools. Such collaboration greatly accelerates the time to market and quality of applications.

CHAPTER 15

Getting It All Together

Now that we have gone through some of the essential building blocks of mobile Internet applications and feel prepared to face the world, it is appropriate to examine some of the related issues. While this book is almost exclusively dedicated to technical issues, the developer still faces many difficult business- and logistics-related challenges. This situation is even more complicated, however, because many of the technical decisions that need to be made are closely related to the choice of business model and other surrounding factors. For start-up businesses, this situation is natural because the borders between technology and business activities are very fuzzy, but some of the software powerhouses might find it very difficult. Other things to consider are how to actually get the application into the hands of users and how to keep them interested. In this chapter, we will touch upon these huge topics and find some general guidelines for developers to use. We will first look at some of the key success factors on the business side, then look at how you can get the products to consumers. Finally, we will stress some of the key features of the mobile Internet that we should leverage.

Business Aspects

Having the best technology and the most skilled engineers is usually not enough to succeed in any business, and the mobile Internet industry is no different. You must know the roles that the different players have in the market and how to move smartly in order to leverage their knowledge and needs.

The Mobile Internet Industry

The business thinking being taught at universities for decades changed as a result of the Internet. The instant access to information across cities and nations has spurred almost as much innovation on the business side as on the technical side. Companies give away goods free just to access new customers, and companies fight about what we commonly call "owning the customer." Getting people to pay for the services has proven very difficult, and customers are notoriously unfaithful (switching to a new site if the terms of the old one do not fit).

The mobile communications industry has (since its major breakthrough in the early 1990s) operated under totally different models. The mobile operator is the main interface toward consumers and sells subscriptions to the services. In these service packages, the user also usually gets a phone that is subsidized by the operator in return for a longer subscription commitment. The idea is that the subscriber should pay little when purchasing the phone and the subscription but then spend lots of money during the time of the contract. In other words, phones are sometimes sold for as little as $1 (or whatever currency is used in the country in question), which is just a symbolic fee that minimizes the barrier of entry. In the late 1990s, the number of users who bought prepaid subscriptions started to rise. With prepaid subscriptions, the operator discount on the handset is usually less, and the phone is loaded with money that can be spent on calls corresponding to that amount. This situation commonly leads to less-faithful subscribers, but people who have lower budgets are especially more likely to enjoy the lack of commitment of a monthly fee. This situation usually leads to a higher level of churn, which is the percentage of subscribers that change to a competing operator. The operator otherwise has a strong position toward the customer and can greatly affect the users' behaviors.

When the Internet and the mobile communications worlds come together in the mobile Internet, roles will change. Companies will have to evolve in order to succeed (and probably in order to survive, as well). Software companies, Web design houses, network companies, telecommunications operators, *Internet Service Providers* (ISPs), and many, many others are now in the same court and are trying to get as many pieces as possible from the available revenue "cake." The user will be offered a large number of applications and services, each consisting of a number of parts (such as wireless access, Internet access, service capabilities, and so on).

For the mobile network operator, this change is huge, and many operators will have to work hard in order to find new roles. Figure 15.1 shows how the revenue that used to completely end up in the hands of the operator now potentially will be distributed among a number of players. The operator might assume one or several of these roles in order to advance in the value chain and to increase the

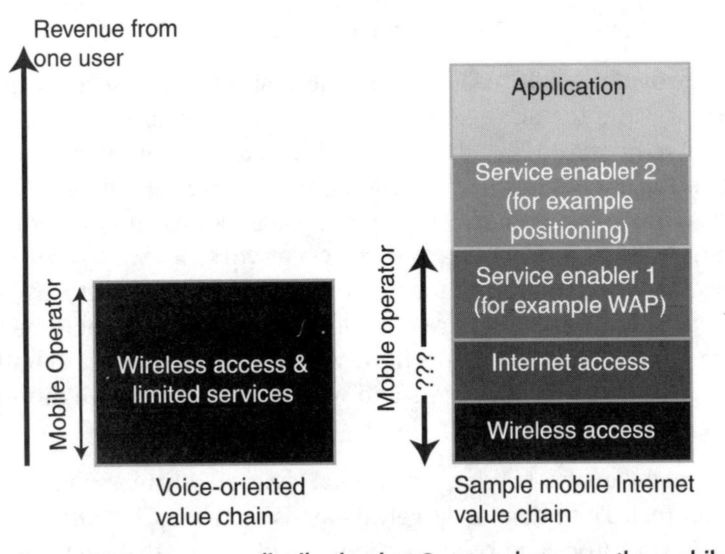

Figure 15.1 Revenue distribution in 2G networks versus the mobile Internet

revenue per subscriber. This method sounds like an obvious choice, but the changes that an operator needs to make in order to become an ISP, for instance, are often substantial (and the chosen business case needs to be investigated closely). In the figure, the operator is viewed mostly as a so-called bit-pipe (in other words, someone who enables the transportation of bits over the air and over other networks). This role is one that will exist in the mobile Internet value chain, as well. Some operators are expected to adapt via the Pac-Man principle: acquiring other operators around the world in order to create a global mobile network where selling the wireless access is the main focus.

One interesting aspect of Figure 15.1 is what happens when an operator successfully climbs in the value chain but keeps the bit-pipe role. Someone who is both an ISP and a bit-pipe can decide to give the ISP access away for free to those who use the network, in order to bring an ISP-only competitor out of business. The same reasoning applies to the operator who gets into the service provider role (for example, by getting a top-class service network). The operator who offers both wireless access (bit-pipe) and services can dramatically reduce the price of the wireless access and charge for the attractive services, or maybe the opposite will happen (charging for services is sometimes difficult). Some operators are expected to clearly separate the service/ISP part from the network access provider part (and in some cases, even make them separate companies). The service/ISP part can then buy air time (wireless access) from any operator. In a country where there are several networks, it is a significant advantage for the service provider to be independent of the bearer being used. Everyone in a family should be able to share photographs, chat, or play games together—even if the underlying bearer is different.

The service provider's role will be very interesting, because it is the place where most of the interesting services are served (but probably the hardest part for which to charge). The advent of the layered 3G architecture opens new opportunities for companies such as Yahoo! and *America Online* (AOL) that are traditional ISPs and content providers. These kinds of companies could start offering mobile services via a service network, as well. This situation would provide an easy migration for many desktop PC users who are unfamiliar with the mobile Internet, where the same e-mail address could be used as well as other applications. The concept of serving similar content to both fixed-Internet (including broadband) users and wireless users but with different formatting is very appealing.

There is no straight answer to what the optimal choice of business model is within the mobile Internet. This situation largely depends on the market and on the competition. While operators and other service providers are struggling to find the right way to approach this emerging market, applications developers are in an easier position. Regardless of who offers services to the subscriber, be it Vodaphone, AOL, or someone else, they still need applications and good content. The applications developer who can develop products that generate traffic the users are willing to pay for or that become a deciding factor in choosing subscriptions will experience the nicer aspects of the supply-and-demand principle. That said, we do not mean that all operators are prepared to pay for the application.

To Charge or Not to Charge?

The first issue surrounding making money from the applications is how the agreement with the operator will look. Some operators offer revenue sharing while others see it as such a privilege to be on their portal that they will not pay a dime to the majority of the application developers. Although some might hesitate to share revenue on traffic volume, the charging capabilities that the service network introduces give a new degree of freedom. By using Jalda (www.jalda.com), the user can be charged on a per-session or even per-click basis, which limits the need for operator involvement. Charging is a very tricky issue, and many heated discussions are expected between large operators and software companies/content providers. The operator argues that the developer should appreciate (read: pay for) the value of being exposed to millions of mobile users. The developer, on the other hand, claims that the operator should understand the competitive advantage of being able to offer this great application. Some developers even have an existing customer base on the fixed Internet that now will use the operator's networks. Discussions involving who should pay whom can sometimes end in an ego competition where each party sees the result as a measure of how important they are. By setting a clear strategy in advance by investigating the operator's policy and evaluating a number of possible business cases, you can mostly avoid this situation.

The question of how much to charge for a service is even harder. Japanese I-mode users have found some services appealing enough to start charging for them. This situation all comes down to how badly the user wants the service and whether it is available elsewhere. The most common way of introducing a new service is to start offering it for free and then add a charge for premium services. One example is to offer a Yellow Pages directory service and later charge for the premium service where positioning is included. In Sweden, the operator Telia offers this kind of service where users can get information about the closest pharmacy, gas station, restaurant, and so on and charges for this service. The operator here has a significant advantage because it already has a billing relationship with the user and can easily add another charge to the phone bill. This solution is also very convenient and secure for the user, who does not have to give away any personal information or credit card numbers in order to make the charging work. The user perception is important here, because people tend to trust the things that they are used to seeing working. When you introduce a new security mechanism, the users not only have to be convinced logically that it works—but more importantly, they need to get a warm, fuzzy feeling that things are under control. The developer can benefit from this situation by working with operators and by leveraging the operator's charging infrastructure.

The difficulty with making users buy wireless applications (some of which are related to logistic issues that we will describe later in this chapter) is likely to spur the development of new business models. One model is to give away the actual application and then charge the user a subscription fee as long as he or she is using it. In any case, getting revenue will be a tough issue for many developers. The Internet has gotten us all used to getting more and more for free, and every user only pays for a few sites (if any). The advertising model of the Internet is much more difficult to implement in the mobile Internet due to the device limitations. In addition, mobile Internet devices will be the most personal Internet access devices that we will have, and every piece of unwanted content is easily interpreted as trespassing on our personal premises.

Be Fast and Be Loud

A venture capitalist-friend of mine has a saying: "If no one else is doing what you are doing, it's probably not a good idea."

After working with developers of mobile Internet applications for more than 18 months, I could not agree more. There are extremely few companies that make products that are totally different from all others. For some of the most common applications, such as *Hypertext Markup Language* (HTML) to *Wireless Markup Language* (WML) conversion and *Virtual Private Network* (VPN) networking, there are tons of them. For most companies, this situation is nothing to panic about—because it at least proves that others believe in the idea, as well. The conclusion is that everyone has to be prepared for head-on competition. The issue now comes down to the question, "Who can be the first and the best?" The latter does not always matter, and often the one that is fast enough and that makes the most noise wins. The noise includes making partnerships, participating at conferences, hiring well-known people, and being active in standardization bodies. The financial markets are often better marketing channels than all of the billboards in the world. If analysts see a hugely interesting pre-IPO company instead of a laboratory of crazy scientists, the press coverage will be substantial and make the partnering much easier.

Being first is not always the most important thing, but being fast is. The first prototype should be developed in a very short time so that the business side can start to work on strategies and partnering while the technical side develops the product in greater detail. Fast prototype development also enables you to test the different features early. There is no better way to learn about the development of wireless applications than to sit in a wireless test lab for an entire day and see your own application being stressed by the different properties. Yes, this book tells you why there will be interruptions and some guidelines on how to avoid the problem, but the feeling of it emerges when applied to something that your team has created. An early prototype is also vital for those who are thinking of having the software preinstalled on shipped handsets. As we described in Chapter 10, "Mobile Internet Devices," the development cycle for mobile Internet devices is fairly long, and the more the application affects the platform, the earlier the manufacturer needs to know about it.

Because there will definitely be some tough fights regarding the "sweet spots" of the mobile Internet market, it is good to keep an eye on the competition. Many of the things that we mentioned previously give that automatically, such as participating in standardization and at trade shows, but other sources such as *Venture Capitalists* (VCs) and the Web are also useful.

Do Not Get into Fights That You Cannot Win

All too many times I have seen applications that are incredibly advanced and that look great at a glance, only to find out what the company is up against on the business side. The most common mistake is for a startup to develop something that has also been developed by a number of companies such as Microsoft, IBM, and Lucent. In this world, David very seldom beats Goliath. If the solution is truly superior, maybe big Goliath brings out his fat wallet and acquires it. One should not despair, however, because big companies sometimes have difficulties with getting smaller products such as applications all the way to the market. With several hundreds of billions of sales, a chat application might not even be commercialized. These big dogs often show lots of concepts but do not always commercialize them. The area where it is most likely that you are going to find competition of this caliber is in the middleware and enablers area—items that facilitate the creation of other applications.

The Internet economy has been fueled to a large degree by money from the venture capitalists that kick-start thousands of startups every year. For every company that is selected by a VC, many are turned down. There are, of course, many parameters that we use when deciding whether or not to accept a company, but an important one is patents. Patents or other indications of a proprietary technology are viewed as a good barrier of entry against the competition. Because there is likely to be fierce competition, a number of nice patents always trigger the interest of all sorts of financial people. The only problem is that everyone else dislikes the ones that have proprietary technologies. Who wants you to succeed if your success means that they have to pay you royalties? Participants in the standardization bodies often use specialists to make sure that a proposed standard uses as few patents as possible. V.42bis is an example of a technology that could have had a much wider spread if it were not for the licensing conditions.

Even harder than getting patents into the standard is trying to develop a better solution than what is already available in a commonly accepted standard. Developing a superior short-range radio technology is probably tough because of the competition from Bluetooth that the other six billion humans on Earth are more likely to be using in a few years. Before creating the superior voice markup language, you should at least know what existing standardization bodies are working on and when they expect to be done with their work. We see again the benefit of having people who work in the standardization bodies help with setting the strategies accordingly.

Get the Right Partners

For anyone who wants to take on strong opponents or maybe even affect standardization, partnering is a key success factor. There are, of course, many candidate partners—and the need and the situation decide what is appropriate. Here, we list some potential partner categories along with the potential partnering benefits for applications developers:

Operators. The operator in many cases has a very strong position in that it has a relationship with the subscribers. This situation can potentially give the developer a prime placement in the operator's portal or can enable charging for services via the phone bill. In addition, operators often look at what their competitors in other countries are offering in order to find new and exciting applications. In other words, a successful relationship with one operator often leads to many more. The developer should therefore consider the situation carefully before giving anyone exclusive access to the application.

Device manufacturers. A terrific way to gain exposure to customers is to have the application preinstalled into a category of devices. Even if many devices will make it possible for users to download the applications that they like, there is nothing like having it built into the product that they buy. Device manufacturers often divide applications into three categories:

1. Those that are preinstalled in the device.
2. Those that are included on the installation CD-ROM (shorter lead time).
3. Applications that users can download and then install.

In other words, even if the device is close to entering production, the developer can still access the attached installation CD-ROM. In addition, the device manufacturers have valuable insights into upcoming devices and trends. These can sometimes be shared with developers under non-disclosure agreements.

Operating systems vendors. While the device manufacturers decide what they will include in their devices, each operating system also often has its own set of core applications. While most of the companies that make the operating systems also develop applications for it, they are mostly open to including new killer applications. Most of the third-party applications that are currently included by default in the existing operating systems have first shown their value as standalone (sometimes shareware/freeware) products. When it is obvious that the users are installing a new mail client although one is included, there is a good chance that the *operating system* (OS) vendor wants it included in the next release.

Network infrastructure vendors. Just like device manufacturers, the infrastructure vendors have very valuable information about upcoming technologies and time frames. This information is otherwise hard to separate from the media hype that surrounds it. These companies usually sell entire solutions, including terminals and applications, but they rarely make all of these products themselves. Most of them need third-party applications to ensure that their customers (the operators) have attractive content to offer with the networks. The agreements with the developers differ and are usually decided on a case-by-case basis. For applications that are strategic to the vendor in question, it might also be possible to gain some deeper cooperation. Because the aim is for them to sell more infrastructure, the applications that are available early in order to showcase new technologies are always interesting and pose a great opportunity for applications developers.

Venture capitalists and other financial backers. Regardless of whether your company is well established or is starting from scratch, you should know the difference between different kinds of funding. One of the great benefits of having the right financial backers is the industry network. Some VCs have specialized in the wireless business and consequently have the right connections to other potential partners.

Simple Is Often Good

In the drive toward developing a product that will change the world in every possible way, some creators of mobile Internet applications tend to make things too complicated. Again, VCs that want to see something clearly unique in the ideas that they fund might be causing this situation. This reasoning is often correct, although there is also a significant advantage in creating simple applications. *Simple* here refers to software or content that is implemented and presented by using existing tools and devices. An example is a WAP directory service or a multiplayer game that is written in Java. None of these require any proprietary technologies, and you can implement them by using available standards.

You can quickly introduce the simple applications on available networks, and the one who launches the service does not have to consider whether this technology will hold. If you develop a complicated platform, you need to convince everyone that this path is unique and is not a path of the competition. A good example of a simple application is the soccer manager game developed by PicoFun (www.picofun.com) where a simple WAP game emerged in early 2000. Many operators saw this product as something new and exciting and that did not require them to make any strategic decisions. For PicoFun, this situation was a way of working closely with operators. An example of the opposite approach is the development of complex multimedia formats. This development would require the operator to make decisions on what formats and platforms to support in the future (a very complex and risky decision). At this point, we dearly need to take on the actions in the previous section of this book: being fast and loud and having the right partners.

Another advantage of making simple applications that adhere to existing standards is that you develop fewer enemies. If you develop an application that uses the UMTS *Application Programming Interfaces* (APIs) specified by 3GPP, all of the infrastructure and device manufacturers will like it because it will showcase their new products in a positive way. In other words, the application also has the chance to become a common best practice at conferences and workshops.

Often, a successful developer starts by doing something that is simple enough to catch people's attention and to enable partnering to get started. As a result, the company builds a good foundation in order to expand into whatever area it desires.

Make Things that Appeal to People

While this action might sound obvious, there are way too few developers who create products that are human centric instead of technology centric. People have always enjoyed products that either make things more convenient or that entertain them. While this factor should be a central issue before starting development, you should also revisit it several times during the development process. The compromises that you make along the way can sometimes affect the functionality and usability of the application.

Again, the operator here plays an important role by wanting not only to please existing subscribers, but also to attract new ones. Anything that will make people say "Wow!" is likely to give a competitive advantage and a potential for gaining a market share. A common mistake regarding otherwise appealing applications is to make the initial acquaintance too cumbersome for the user. The number of barriers through which a user has to pass should be minimized, which includes massive navigation as well as initial passwords in order to get started. For every navigational step that you take, you will lose some users—and once user input is needed, even more drop out. An attractive application should tease the first-time user with things that are ridiculously easy to use. Once people have gotten addicted, they are less likely to drop out if they are presented with a registration window.

Distribution and Maintenance

Everyone is talking about the convergence of the Internet, the software industry, and the telecommunications world, which creates some new questions (for instance, "Should mobile Internet applications be distributed?"). Software companies have gotten used to selling the majority of desktop PC applications via retailers by packaging it into a box. Retailers can then be anything from online e-commerce sites to brick-and-mortar warehouses. The goods sold are tangible, however—you can walk into a store and buy a box containing the game Quake 3 Arena, and it will contain documentation and a CD-ROM. The question now arises as to whether this method is the way that mobile Internet applications are to be sold as well? For WAP applications, the need for distribution is not there, so here we are talking about terminal-based applications that execute on the actual device.

Some applications will be exclusively distributed via the Internet, where the user starts a session by clicking a link that downloads the code and then erases it once the execution finishes. Others will be downloadable as shareware to be installed on the device in question. The charge can then be made on a subscription basis, as we mentioned previously.

The devices and their installation CD-ROMs are, as we mentioned in the previous section, one of the most valuable channels for accessing the consumers. From the desktop PC world, we have learned that many users just stick with whatever is included in the bundle that the retailer offers. While this method is by far the most difficult distribution method, the option of being included on the installation CD-ROM is a bit more realistic for the general developer.

The network operator will also have a package of applications available to new and existing subscribers. For server-based applications (WAP), this situation means being a part of the operator portal and maybe even getting a prime position in the portal. For terminal-based applications, some operators will preinstall a few applications while others will include them on the operator's installation CD-ROM.

Because this field is completely new and no one seems to have all of the answers, we are likely to see lots of innovation in the area. For instance, it would be possible to sell the application with a box that contains a Bluetooth device that then installs the application via Bluetooth when you press a button. Bluetooth also facilitates the rapid spread of those freely downloadable applications that are useful. Once a few enthusiasts have downloaded them off the Web, others can get them transferred to their devices over Bluetooth. This method has previously been shown to be a very effective way of getting the application to penetrate a market. Examples include the Palm hand-helds and the Hewlett-Packard scientific calculators. In both cases, infrared is used, but Bluetooth should sometimes make things even easier, assuming that this method of distribution again requires the revenue to be collected by new means. Subscriptions based on usage are one way, and other methods are likely to emerge.

Getting the application out there is only the first step toward handling mobile Internet applications logistics. Ensuring that the users get proper service and maintenance is also essential. Especially those applications that are terminal based can be difficult to maintain. Fortunately, terminal based mostly means that there is a server end as well, and automatic upgrade functionality should be included from day one. You run the risk of ruining the market for a perfectly good application by introducing multiple, incompatible versions at the same time. Here, the cooperation with manufacturers of the devices and adherent applications environments can help with anticipating when new releases are needed.

Know Your Network, the Operator, and the Consumer

Being successful depends very much on being prepared. The developer who has read this book should know a lot about the prerequisites of implementing applications all the way to the consumer. Although most developers do not want to be confined to one or a few markets, the knowledge about individual markets is often valuable. As a result, the developer should know what network is used, what business models the operator has, and what to expect from consumers.

Knowing the network makes it possible to predict when new features will be implemented and what the resulting characteristics are going to be, including service capabilities (Chapter 9, "Application Architectures") and the realistic expectations of bit rates and QoS. Other issues include what the future upgrade path is and how users can bring their handsets into other operators that are located nearby and still use the same services (so-called service roaming). You will find information about the network properties and upgrade paths in chapters 2, 3, and 4.

In a time of rapid change, it is vital to understand the thinking of the operator. Is he or she going to offer applications, or will he or she use an external service provider? This understanding includes knowing what service capabilities will be offered, such as positioning and micropayments. By networking with other developers, you can often find out the operator's general policy toward developers. Some have decided never to share revenues from traffic volume, while others are more flexible. The operator is also the one who decides exactly when a certain feature of a network will be launched (regardless of the delivery dates from the network infrastructure vendors).

Finally, you should know the consumer's needs as much as possible. While the strategies for getting this information are the same as for any consumer product, there are some wireless-specific issues to consider. When approaching a certain user group, you should know both the Internet and wireless penetration level of the consumers. If possible, also know how much money the group is spending on those services and *how* they are accustomed to paying for applications. Some might pay with credit cards while others might pay via the operator's phone bills. Any way, charging in ways that consumers are used to makes the new application less of a transition in user behavior.

In order to be open to worldwide expansion, you should make all of these decisions while keeping these factors in mind. If you choose a certain charging scheme, how much needs to be changed in order to adopt it to other schemes? There are usually fairly simple ways to make an application adapted for local conditions while still keeping the flexibility that is needed for global implementation.

Leverage the Unique Possibilities

After reading this book, many developers run the risk of becoming too focused on the challenges ahead when creating applications for the wireless environment. There are indeed a number of difficulties that they have to overcome, but the opportunities are even bigger. At the last part of this book, I want you to look back and see the amazing opportunities rather than the obstacles. The mobile Internet has some features that make it unique, and the applications for the mobile Internet should leverage these features. Users will still use other means of accessing information, including Internet and televisions, and a crucial success factor is to ensure that mobile Internet applications are built upon the strengths of these features.

With the advent of new technologies, it is always hard to anticipate what the use for it will be in the years to come. There are usually several good ideas, but usually the things that appear after several years of learning are so advanced that no one would have predicted them. When computers first became pervasive, no one could anticipate the wide range of uses that we take for granted today. What *was* known at that time, however, was the properties of computers and their basic features. Computers were fast when calculating and handling large amounts of information. They were flexible, but still the results from operations were always consistent (for example, judgments were not based on feelings but rather on rules). By knowing about these things and understanding the challenges as well as the opportunities, the industry could evolve the computer and its applications tremendously. Next, I describe three important features that I hope will inspire you to not only survive the mobile Internet revolution, but also to make the most of its many advantages.

Personalization

The most successful Internet sites have built a lot of functionality around presenting a personalized view for the user. At logon, the user is welcomed by name and the preferred settings are automatically activated. For e-commerce sites, the profile also includes payment information in order to ensure a simple click-and-buy experience. What was a competitive advantage in the beginning has now grown into a prerequisite for attracting customers (and has become a vital part of any application).

After seeing the importance of personalization for fixed-Internet services, some say that the mobile Internet will be no different. Nothing could be more wrong, however. For the mobile Internet, personalization is even more important and will be one of the key driving forces behind the entire market. Making the applications user-centric as opposed to technology-centric is crucial. Ensuring that the user feels as if the application adapts to his or her needs will not only attract many users, but also more importantly will keep them from switching to other competitive applications.

The devices that people use when they are on the move are more personal than any other Internet access device. In many countries, people have already chosen 2G phones with the design as their number one criteria. Ultimately, we will all have mobile Internet devices that reflect not only our needs, but also our personalities. For some, the device will have a simple but elegant look and will give one-click access to stock-charts and travel information. For others, the curvy exterior and shiny colors are complemented with games, chat applications, and alerts about celebrity activities in the nearby area. The 2G phones took the first step toward personalization with more design-centric devices and interchangeable shells. The next step is the personalized interior. Not only will the functionality and features be different for different target groups, but the software will also differ for different individuals.

Applications have traditionally been categorized into vertical and horizontal segments, depending on their target groups. The vertical applications fulfill specific needs (dispatch handling, medical applications, and so on) for a more narrow target group while the horizontal ones span many segments (mail, chat, and so on). With personalization, this functionality will be taken to a new level with a separation on a personal basis. For me as a user, the applications are either part of my personalized applications or part of the "unpersonalized" ones. The difference is not only that the personalized ones are tailored to my needs and are pumped with my personal preferences, but also that they have an ease of access.

Personalized applications will be available for instant access—just a click away (with whatever input device is used). As an example, for WAP applications, the personalized applications are accessed via the portal that the browser displays. These are (supposedly) the most commonly used applications and are of the highest importance to the user. The operator might force some initial settings and portal content to new subscribers, but those who are driven to success will enable the users' easy personalization of the portal. For open-platform devices, the personalized applications can be the ones that are accessed via icons on the device. As the mobile Internet becomes a natural part of every PDA and communicator, there will be little difference in accessing the datebook and to-do list and the service that gives the weather forecast for the area in which the user is located.

While personalization traditionally has meant tailoring an application to the specific needs and properties of users, the mobile world takes a step farther. The applications will be increasingly location-aware, and the personalization should of course be expanded to include this factor. Some applications are very location-centric and become the core of the entire product, including mapping applications and tools to find the shortest route and the closest restaurant. Other applications use a more discrete personalization based on the current position. A currency exchange application might suggest British pounds as the currency to convertfrom when it detects that you are in England. This location-specific information could then be complemented with the country in which you live. This function could be part of the personal service environment (mentioned in Chapter 9, "Application Architectures") or a part of the device or application profile. In this way, the currency conversion application uses your home country and the country you are currently in as the default choices.

To summarize, we can describe the main parts of the personalization as follows:

- Tailoring the application to the needs and preferences of the user
- Ensuring that the entire device is personalized, both externally and internally (includes easy one-click access to the applications that reflect the personality of the user)
- Making the application both user and location-aware

As we can see, the personalization concept is widely expanded in the mobile Internet and is a vital part of its success.

Always Online

The main reason for implementing GPRS on top of existing 2G networks is the always-online feature. Although it still might take a few seconds to fetch the information, it is a significant step toward making the mobile Internet more user-friendly. The first applications are merely enhanced by it (for example, removing the dial-up process for WAP applications and charging for things other than just time). Developers who start developing for networks where instant connectivity is a natural part find that there is so much more that they can do.

When you always have the network at hand, the applications can be built around the network connectivity in a new way. The wireless applications will be no more complicated to access than the ones that merely access the device itself. The relationship between the client and server parts of the application becomes much tighter, and the server can instantly notify the client about interesting events. When the application also is personalized to react according to the needs of the user, we get closer to applications that really make things easier for us.

As an example, I configure my device to be aware that I like jazz clubs and that I am interested in knowing when good jazz gigs are in town. When I now move around in a city that I am visiting, I know that my device will give me a vibrating alert if there is something going on. The alert can either be dismissed, or I can request more information, such as where the event is, who is playing, and so on. The concert information lets me choose whether I want to listen to a short clip by the band in question or perhaps see a short live video from a similar jamming session in this club. Because I am subscribing to these alerts, I also get special discounts if I want to purchase the tickets or even a CD on the spot. Because the application is actually helping me, I do not mind a small bit of advertising. Perhaps when I select to receive directions to the club, the application will also ask me if I need some food on the way or a parking spot nearby. Now that I have indicated that I am going there, the device can add this feature in order to enhance the level of personalized information even more with alerts about traffic incidents and so on.

The concept in this example can be generalized into a concept of agents or *bots* that operate on your behalf in the device. There are, for instance, price-comparison agents on Web sites that find the best price for buyers of a certain item. These are reactionary and task oriented, however, because you tell them that you want the best price for a TV, and they then look for it. This feature, however, reflects the current status—and you rarely find those that continue to search for a few days and then get back to you. With a bot inside your mobile device (facilitated by an open platform), you can let it know what you are looking for and will know that it will get back to you when something comes up. Instead of the classic request-response user scenario, we move toward getting a device that knows what we need and that only gives a response when there is something to report. A bot could also be told to download and/or synchronize certain information when the traffic load is low (as well as when the cost of transmission is low). Perhaps you want to synchronize your calendar with the home or work PC, both to ensure consistency and to keep a backup of the information. The bot in the device could then find an appropriate time to perform this task when you are not busy doing something else with it.

Mobility

The third key feature of wireless applications that I wanted to emphasize is the core of it all: mobility. While there will undoubtedly be a lot of synergies between applications on wireless and fixed networks, the wireless part should be taken even farther. The mobility is something that the application not only should survive, but also leverage in order to make the application benefit from the independence of fixed access points.

When you are designing applications, the particular use case in question should always be your focal point. With common databases and content for desktop PC applications and mobile applications, there is a risk that only the user interface will be adjusted for the wireless part. This is a big mistake, however, because it reflects the view of wireless applications as a necessary evil rather than the great opportunities they bring. In addition to making the presentation appropriate for the mobile devices, the actual application logic should reflect a mobile thinking. The users will have this application in their pockets or in their bags, which makes it available for the tiniest impulse. The impulse of looking up the meaning of a word might be too small for someone to boot up the desktop PC and find the answer. Whether the user decides to look the word up or not also depends largely on his or her current whereabouts. If you are walking down the street, you need a very strong impulse before you actually look up a PC on which you can find the answer. The mobility of wireless applications gives the users less of a threshold for these kinds of small tasks because the device can be carried around. If you are on a bus and want to find out something, it is easy to do it when the information is one or two clicks away on the mobile device.

Summary

Creating successful wireless applications is not only about the technologies involved. Many other things need to come together, and we must not assign a low priority to the business aspects. Understanding the different players of the mobile Internet and how to work with them is one of the key success factors. Everyone benefits from partnering, and it often makes the development process easier. The distribution of wireless applications can sometimes be a tricky issue. Again, partnering opens many channels and takes a lot of these worries off the shoulders of the developer. Often a simple but appealing application that adheres to standards is more likely to succeed than a more complex product that has many proprietary components.

By sharing my knowledge about mobile Internet components with you, I hope to spur your imagination and lead you to many creative ideas. Hopefully, closing this book means that your imagination is opened for many new and exciting things. Good luck!

Acronym List

2G Second Generation (mobile systems)
3G Third Generation (mobile systems)
3GPP Third Generation Partnership Project
AMPS Advanced Mobile Phone Service
AOL America Online
ARIB Association of Radio In Business
ASP Application Service Provider
ASuS Application Support Servers
ATM Asynchronous Transfer Mode
AuC Authentication Center
BSC Base Station Controller
BTS Base Transceiver Station (base station)
CDC Connected Device Configuration
CDMA Code Division Multiple Access
CDPD Cellular Digital Packet Data
CGI Cell Global Identity
CLDC Connected Limited Device Configurations
CORBA Common Object Request Broker Architecture
CS Circuit-Switched
CS Coding Scheme
D-AMPS Digital AMPS
DECT Digital Enhanced Cordless Telecommunication
DSL Digital Subscriber Line
EDGE Enhanced Datarates for Global Evolution, *or* Enhanced Datarates for GSM and TDMA Evolution

EIR Equipment Identity Registry
E-OTD Enhanced Observed Time Difference
ETSI European Telecommunications Standards Institute
FA Foreign Agent
FCC Federal Communication Commission
FDD Frequency Division Duplex
FDMA Frequency Division Multiple Access
GAP Generic Access Profile
GGSN Gateway GPRS Support Node
GOEP Generic Object Exchange Profile
GPRS General Packet Radio Services
GPS Global Positioning System
GSM Global System for Mobile communication
GSN GPRS Support Node
GTP GPRS Tunneling Protocol
HA Home Agent
HLR Home Location Registry
HSCSD High-Speed Circuit-Switched Data
HTML Hypertext Mark-up Language
IEFT Internet Engineering Task Force
IMT International Mobile Telecommunications
IP Internet Protocol
ISDN Integrated Services Digital Network
ISM Industrial-Scientific-Medical
ISP Internet Service Provider
ITU International Telecommunication Union
IWF Inter Working Function
JVM Java Virtual Machine
kbps kilobits per second
KVM K Virtual Machine
LAN Local Area Network
LDAP Lookup Directory Access Protocol
LLC Logical Link Control
MExE Mobile Execution Environment
MIDP Mobile Information Device Profile
MMI Man-Machine Interface
MOA Mobitex Operators Association
MPC Mobile Positioning Center
MS Mobile Station
MSC Mobile Switching Center
MT Mobile Terminal
OPL Organiser Programming Language
PCN Packet Core Network

PCS Personal Cellular System (the 1900 frequency band)
PCU Packet Control Unit
PDA Personal Digital Assistants
PDC Personal Digital Cellular
PDP Packet Data Protocol
PDSN Packet Data Serving Node
PIM Personal Information Management
PIN Personal Identification Number
PKI Public Key Infrastructure
PPP Point-to-Point Protocol
PQA Palm Query Applications
PS Packet-Switched
PSE Personal Service Environment
QoS Quality of Service
Radius Remote Access Dial-In User Service
RLC Radio Link Control
RNC Radio Network Controller
SCS Service Capability Servers
SDAP Service Discovery Application Profile
SDK Software Development Kits
SGSN Serving GPRS Support Node
SIM Subscriber Identity Module
SMS-C Short Message Service Center
SPP Serial Port Profile
SS7 Signaling System 7
TA Timing Advance
TACS Total Access Communications System
TCP Transmission Control Protocol
TDD Time Division Duplex
TDMA Time Division Multiple Access
TE Terminal Equipment
TLLI Temporary Location Link Identifier
TLS Transmission Layer Security
TOA Time of Arrival
TS Timeslot
UMTS Universal Mobile Telecommunication System
VHE Virtual Home Environment
VLR Visitors Location Registry
WAP Wireless Application Protocol
WCDMA Wideband Code Division Multiple Access
WLAN Wireless Local Area Network
WTLS Wireless Transmission Layer Security

Index

Symbols

A

B

C

E

F

G

H

I

J

K

L

M

N

O

Q

R

S

T

CUSTOMER NOTE: IF THIS BOOK IS ACCOMPANIED BY SOFTWARE, PLEASE READ THE FOLLOWING BEFORE OPENING THE PACKAGE.